Into the Forest

A Woodland Coloring Book
for Teens and Adults

by Dandelion & Lemon

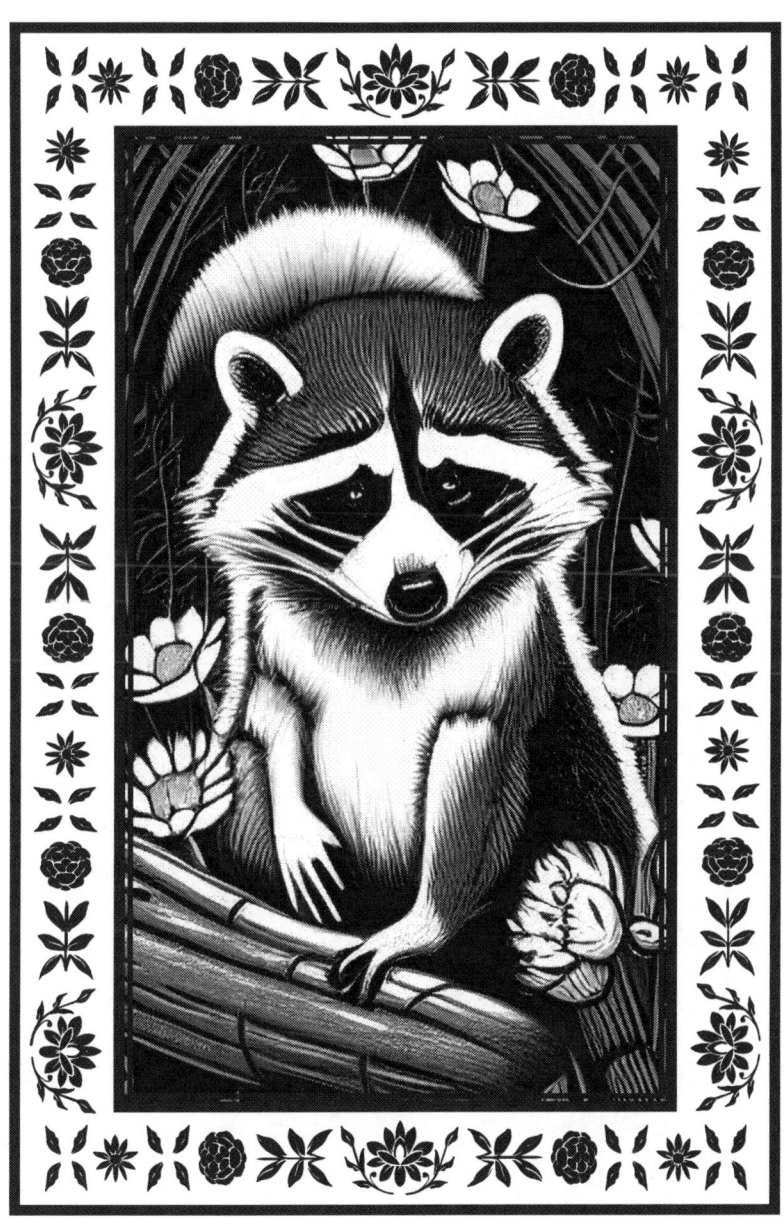

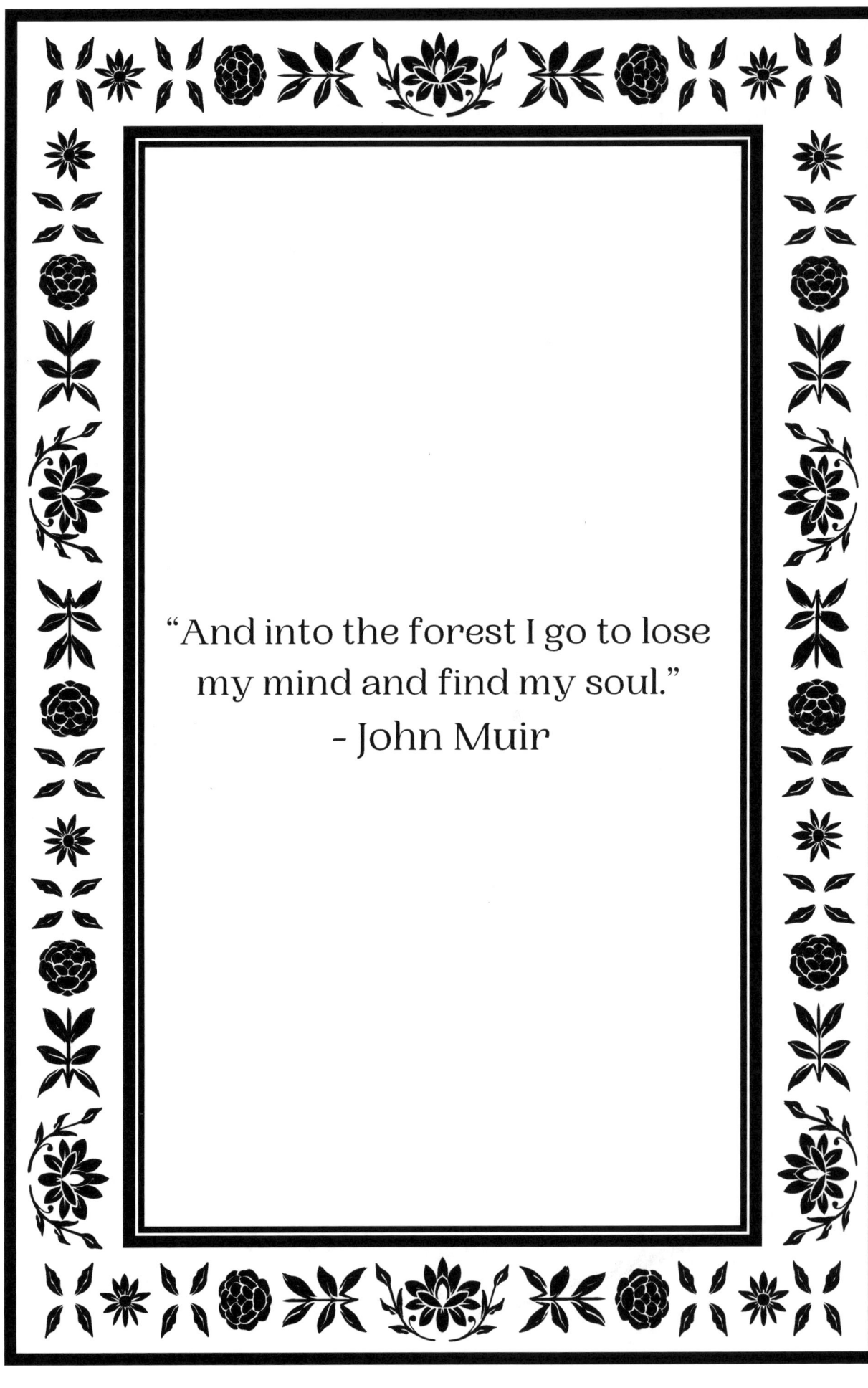

"And into the forest I go to lose
my mind and find my soul."
- John Muir

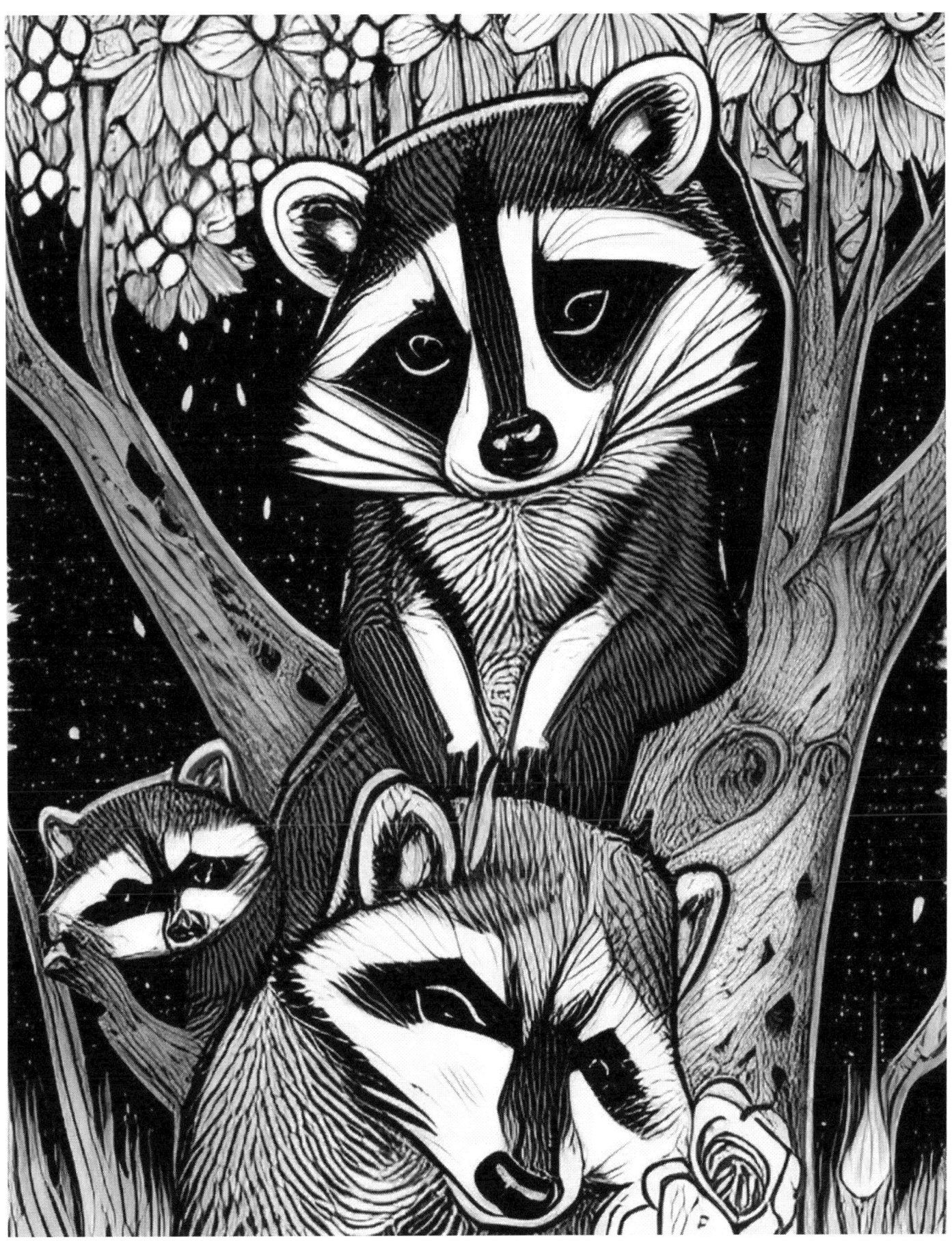

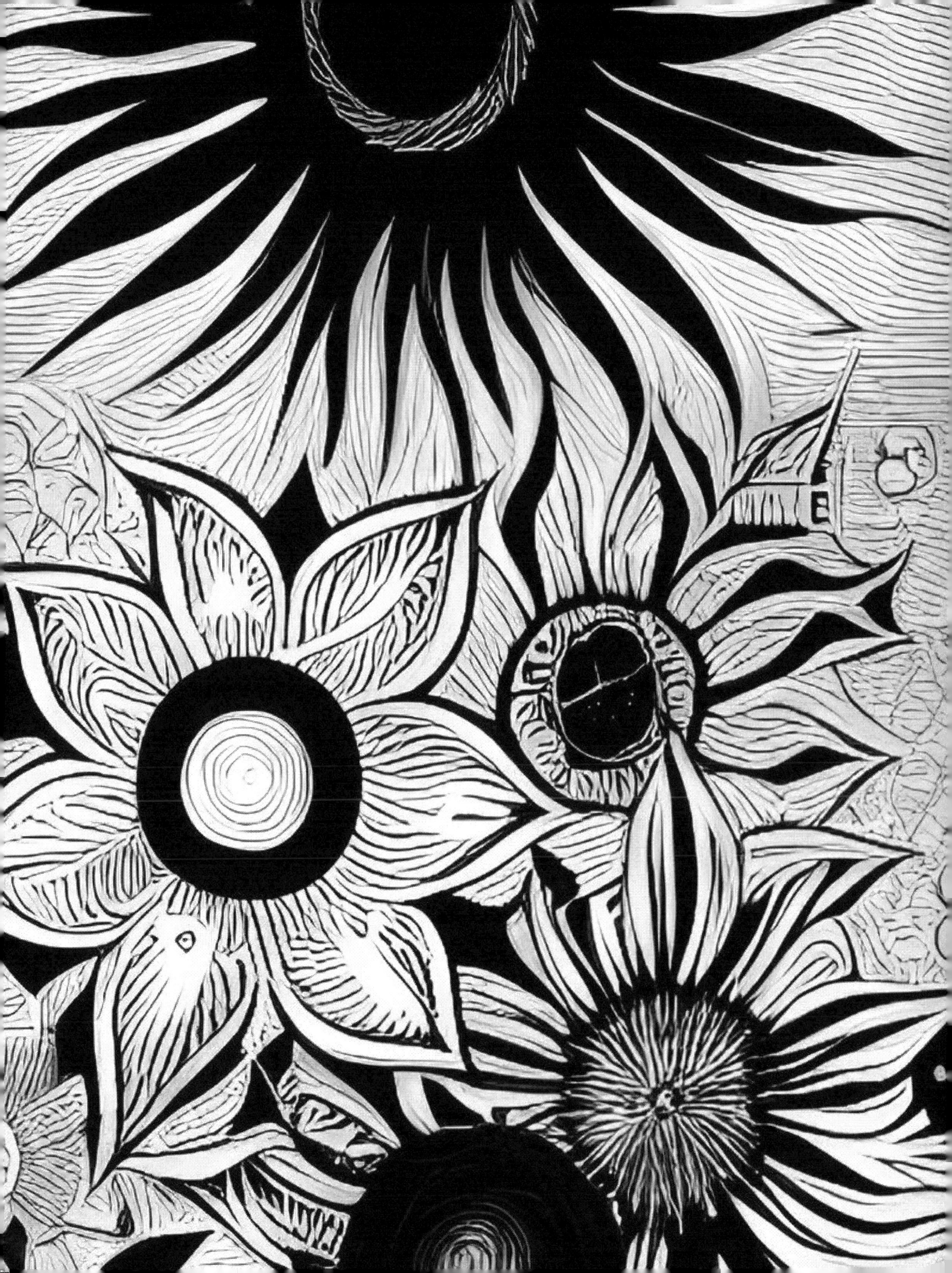

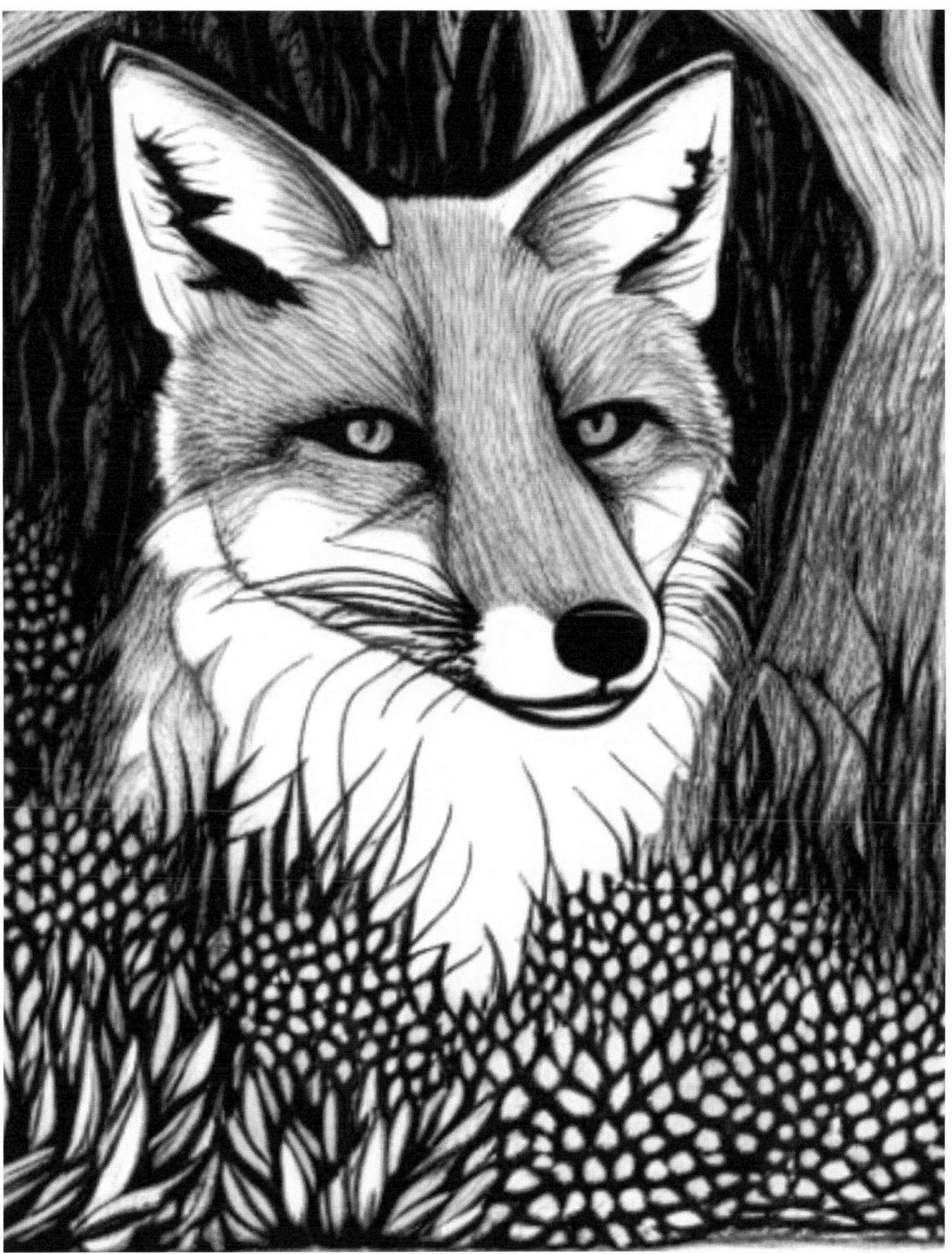

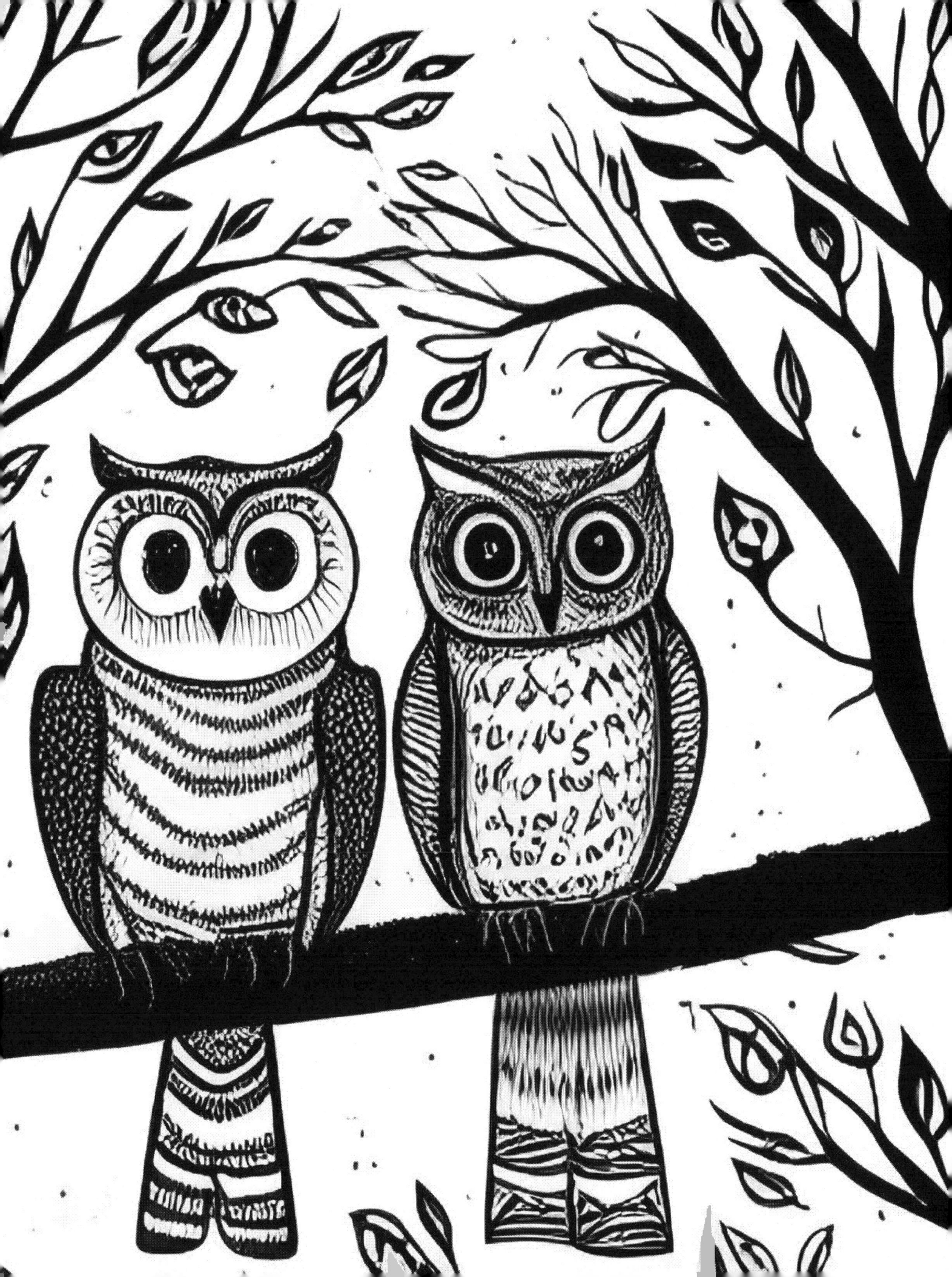

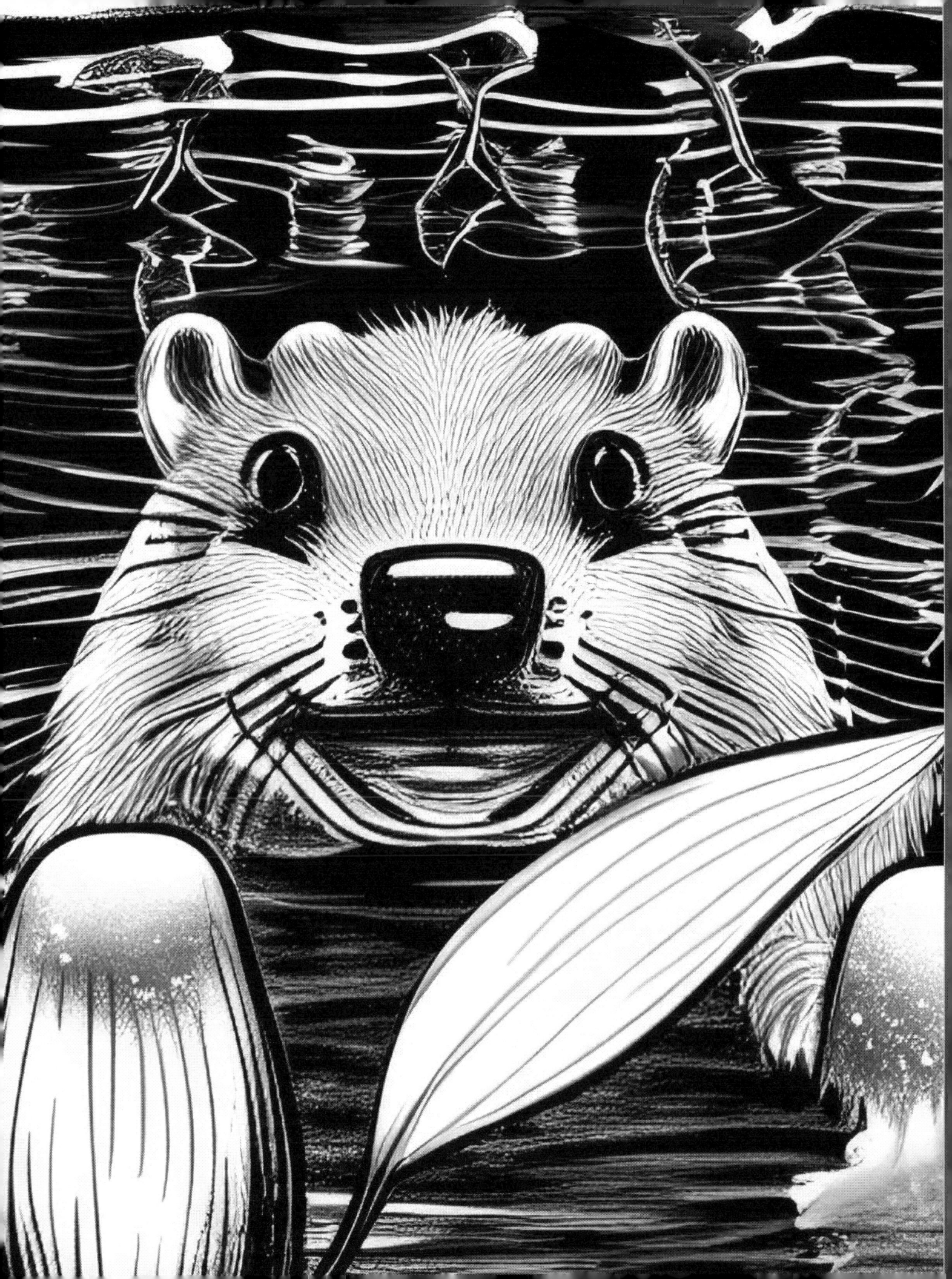

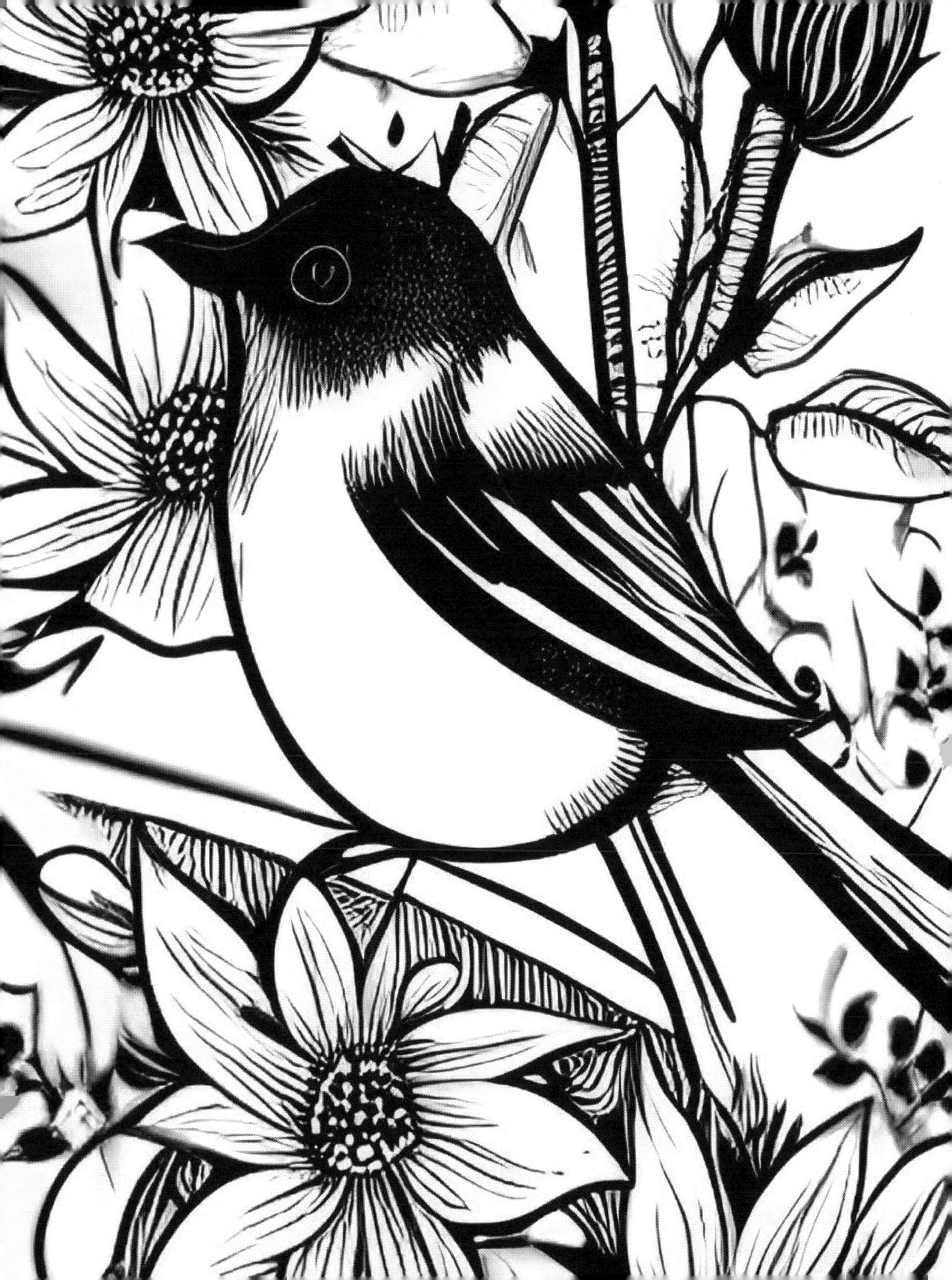

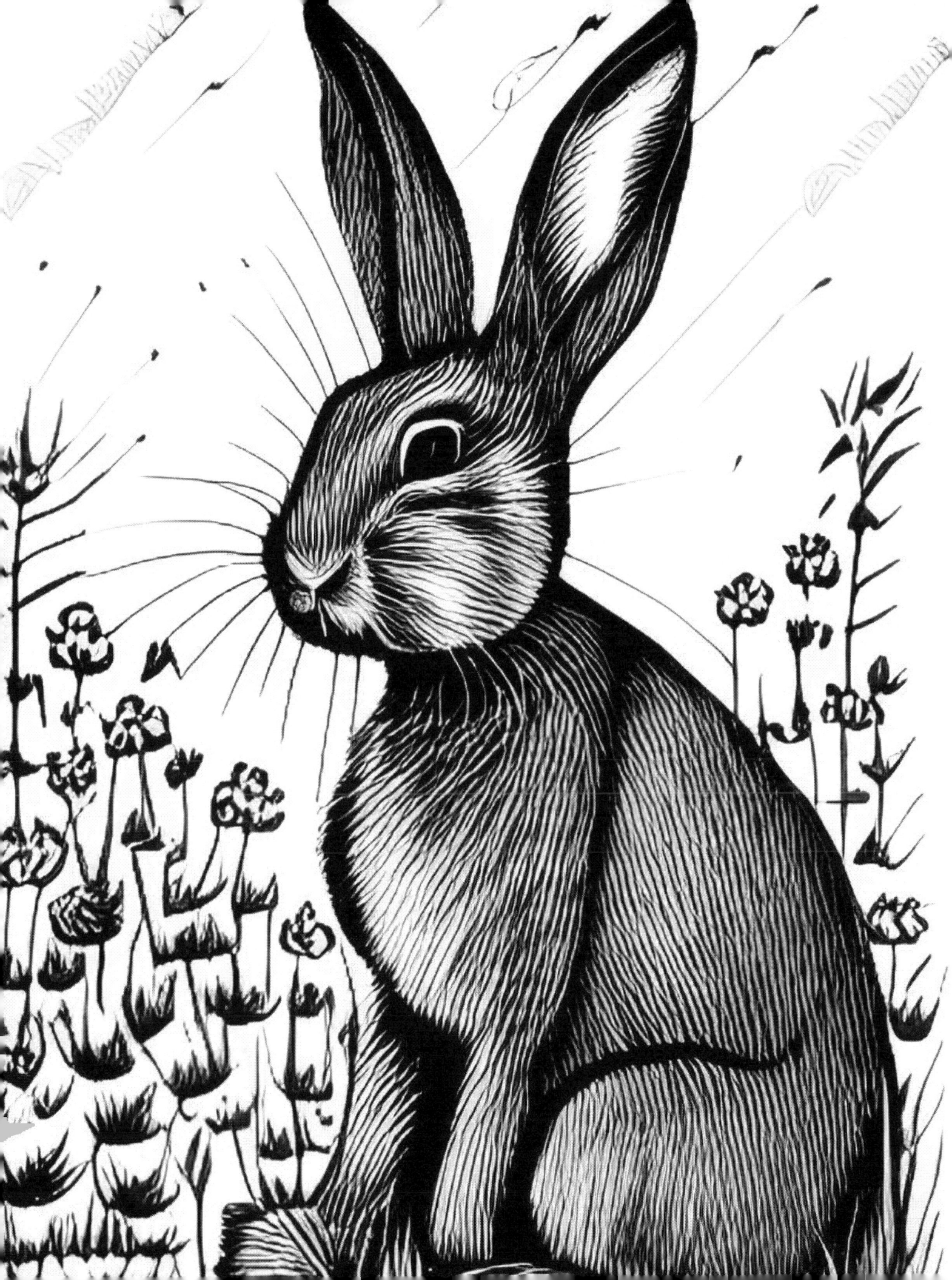

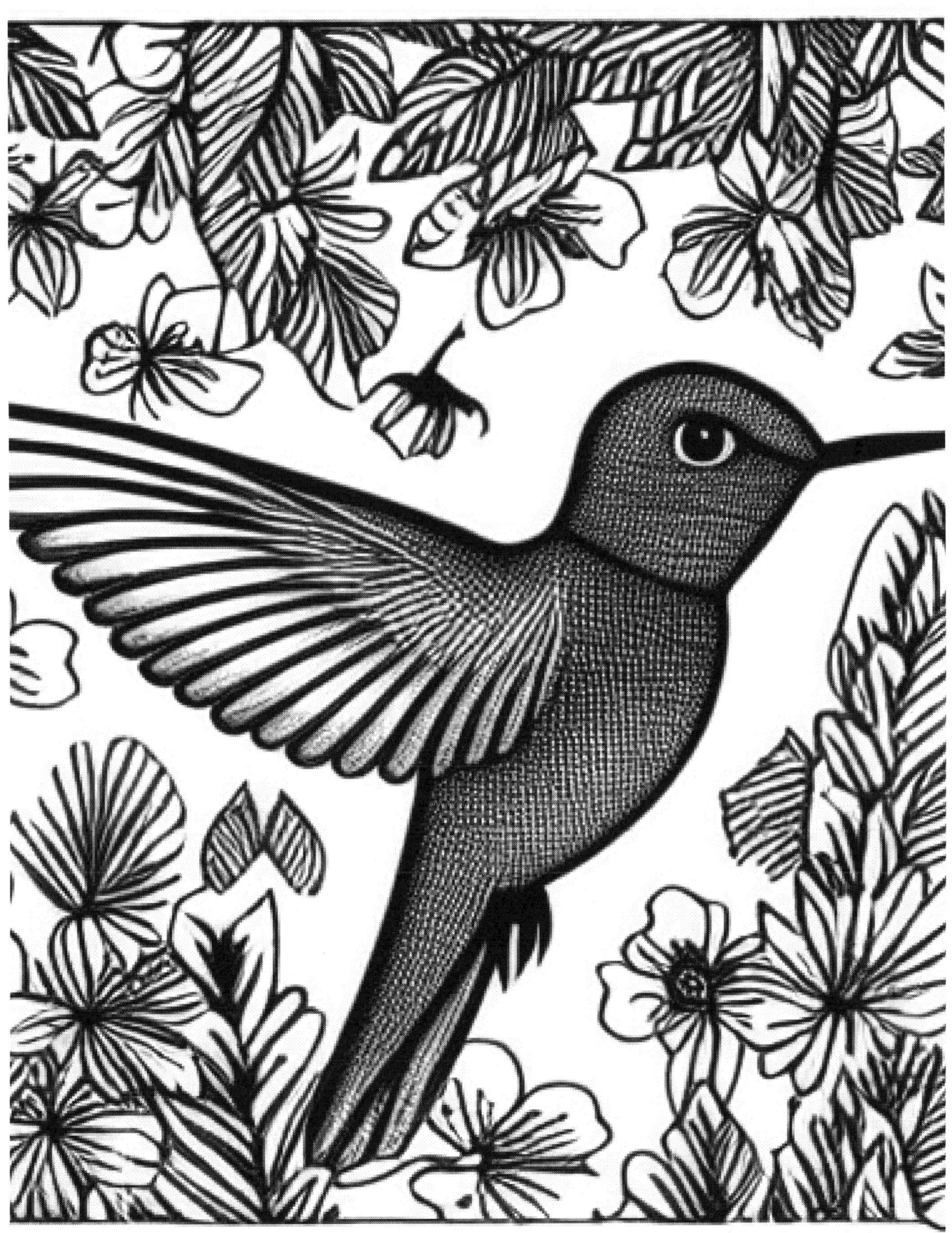

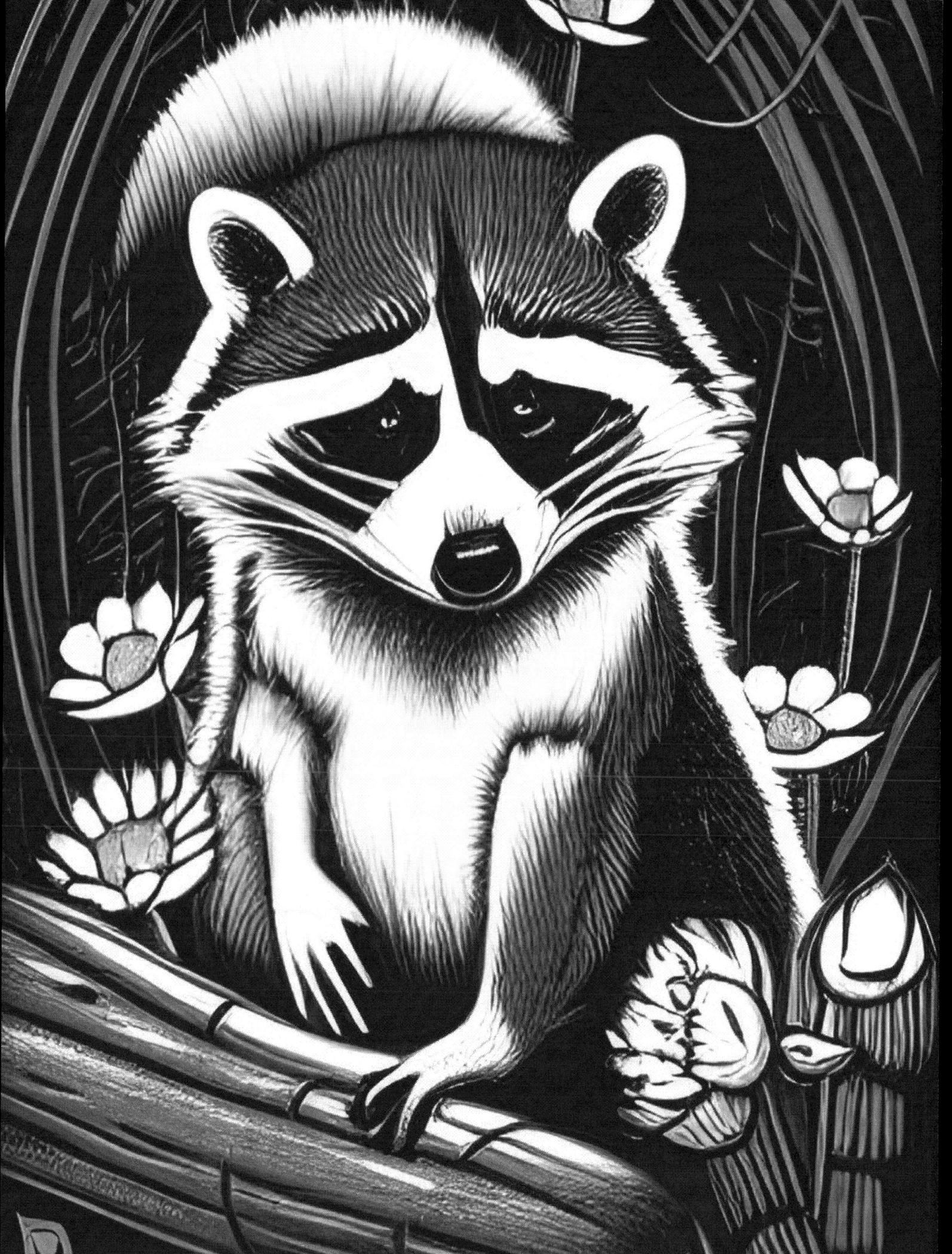

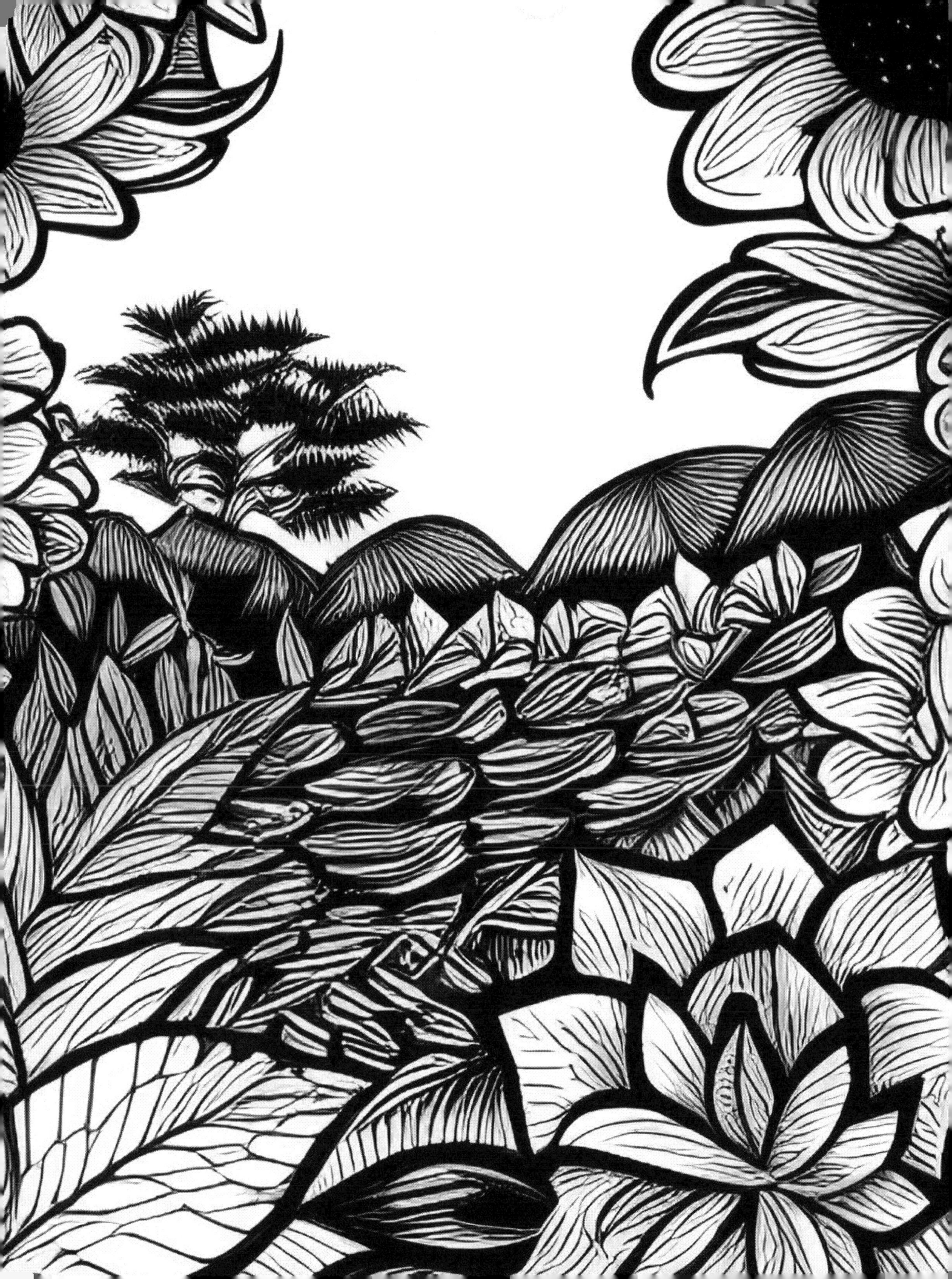

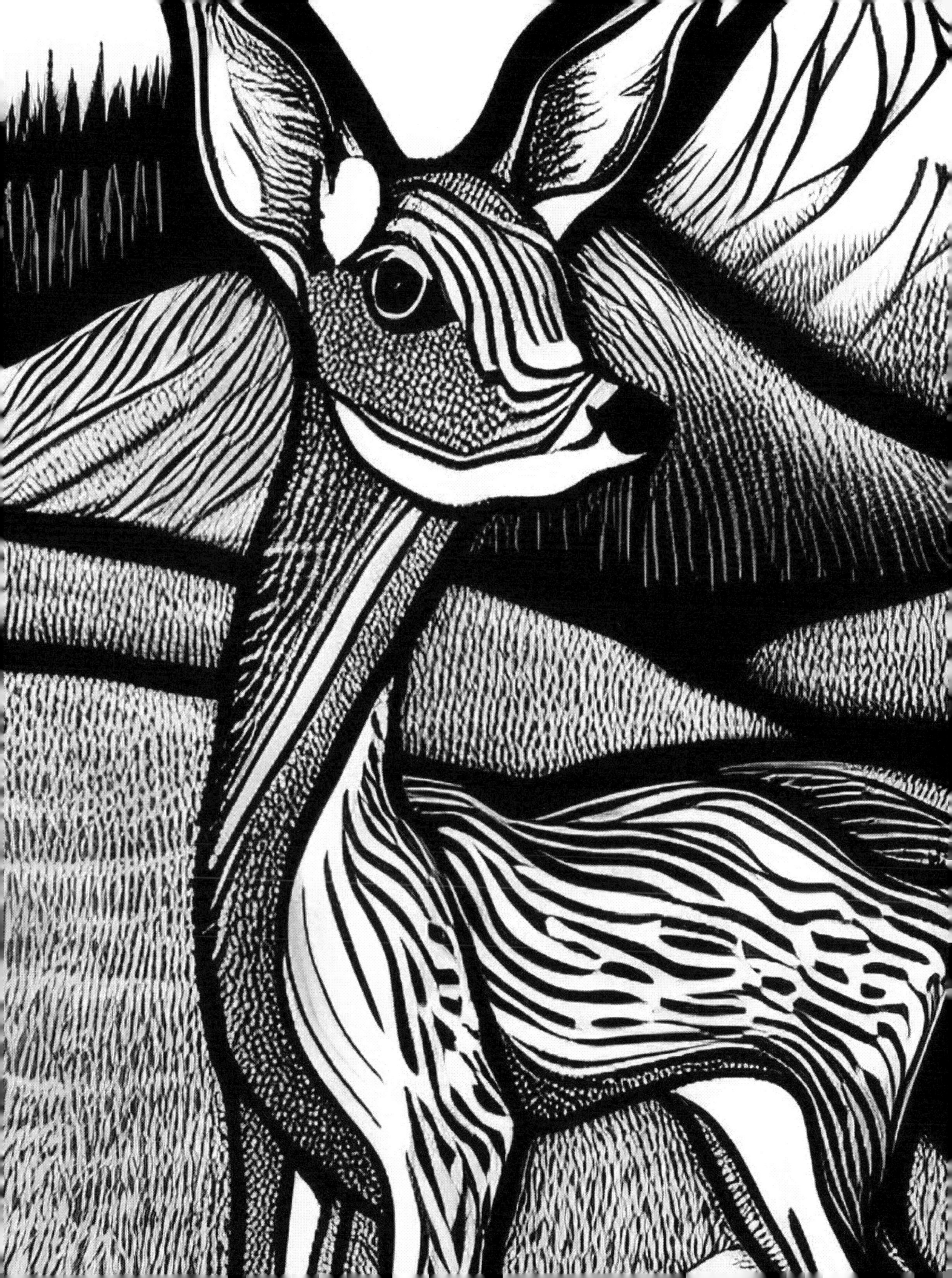

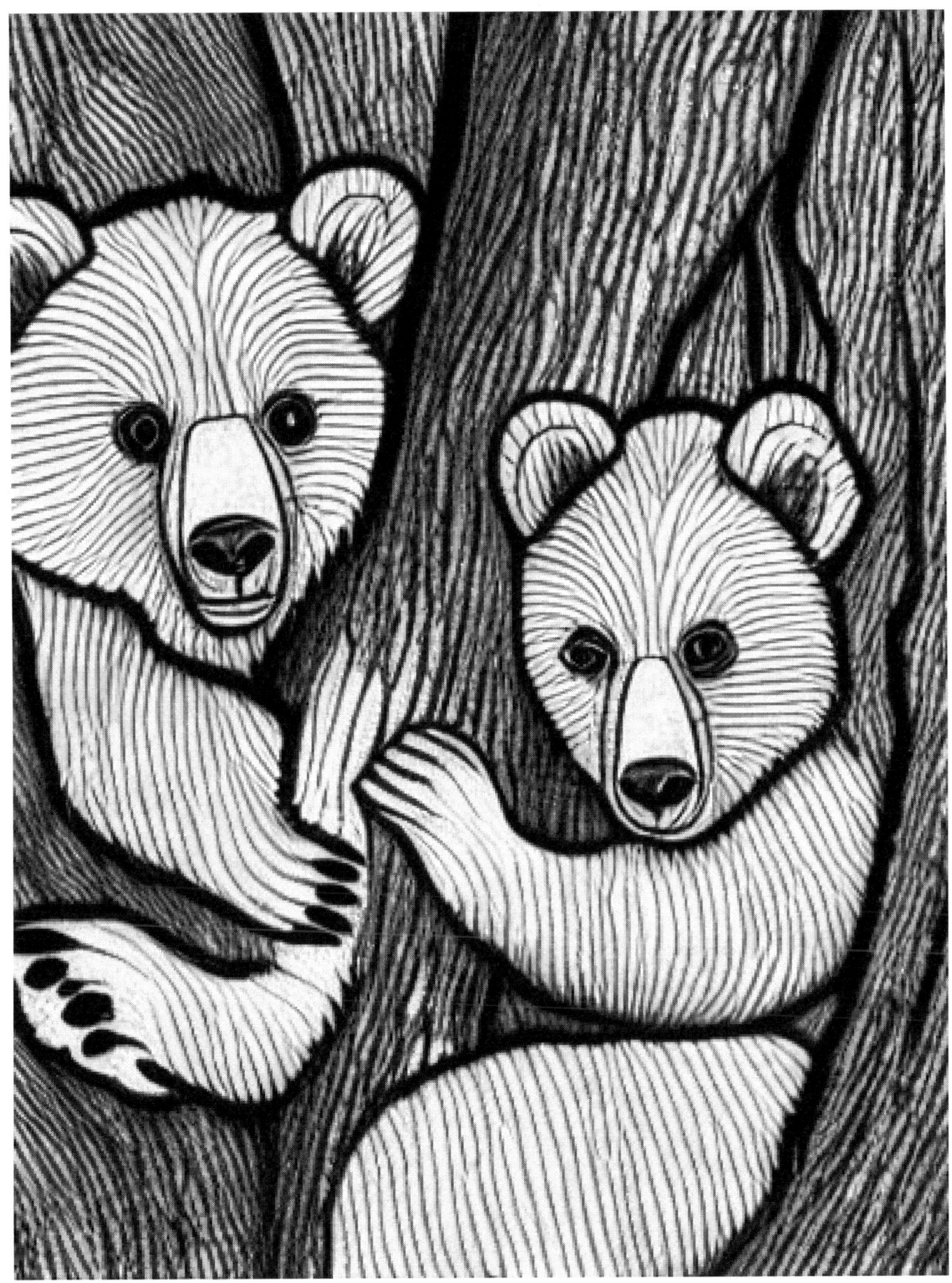

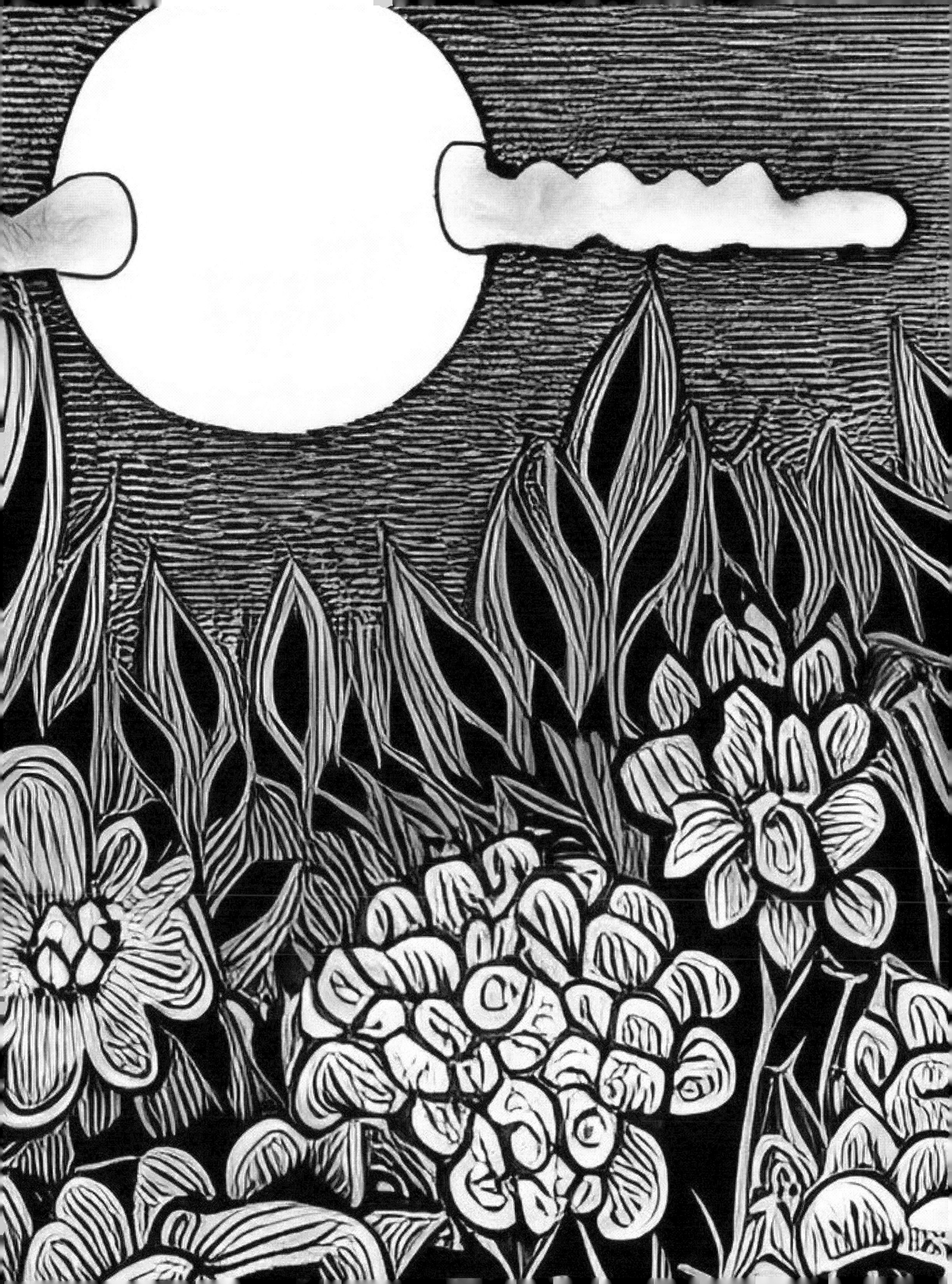

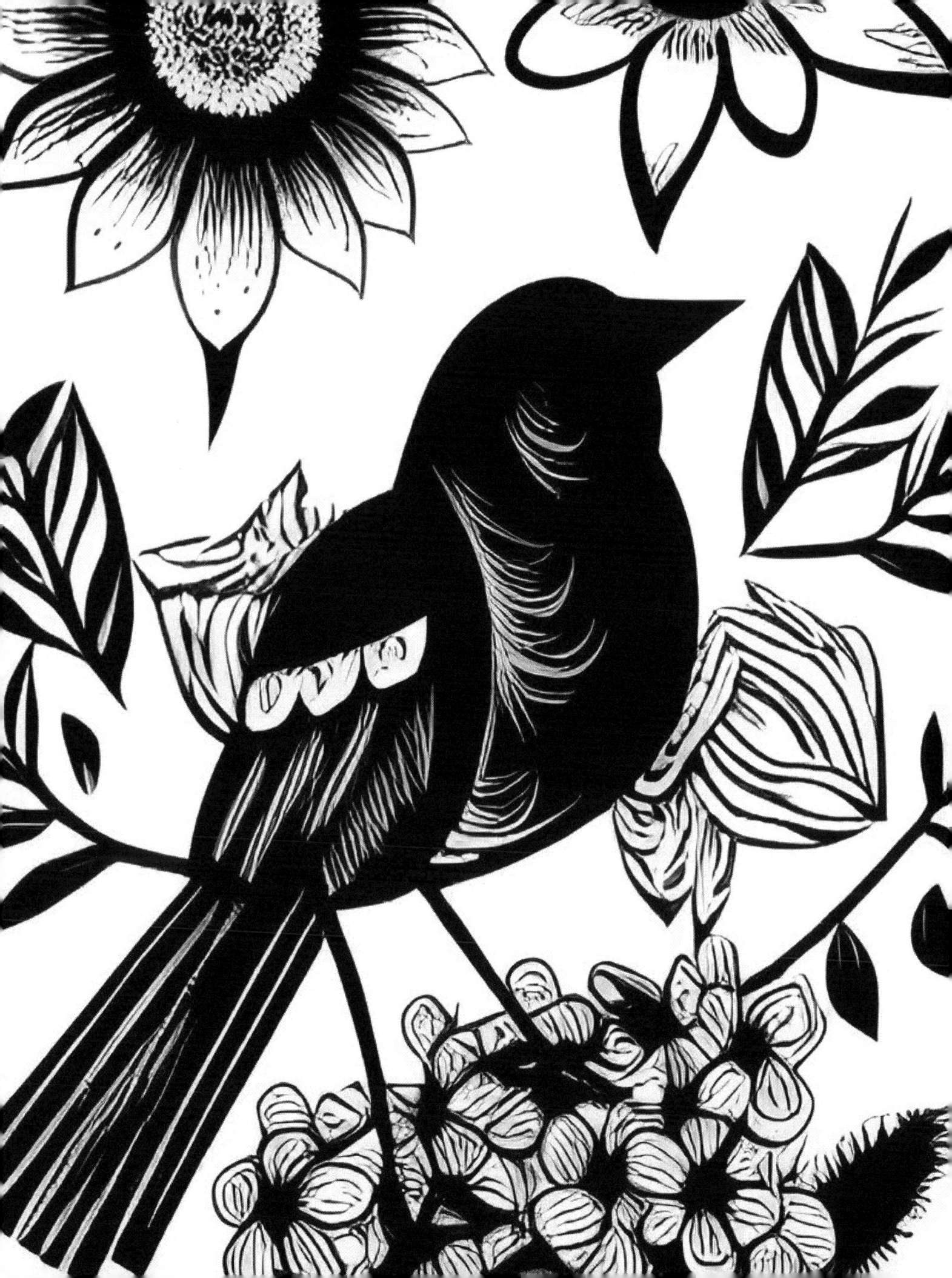

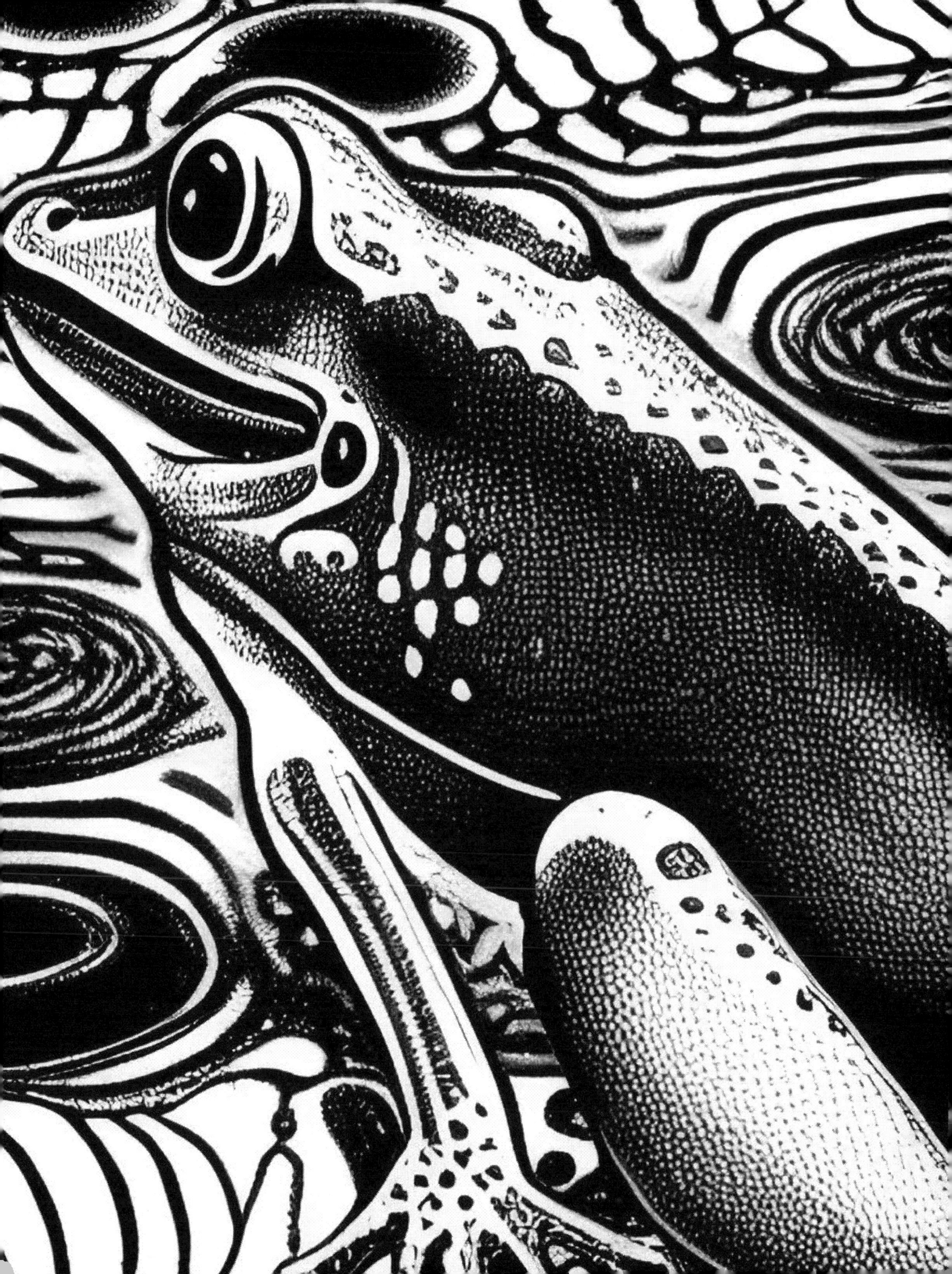

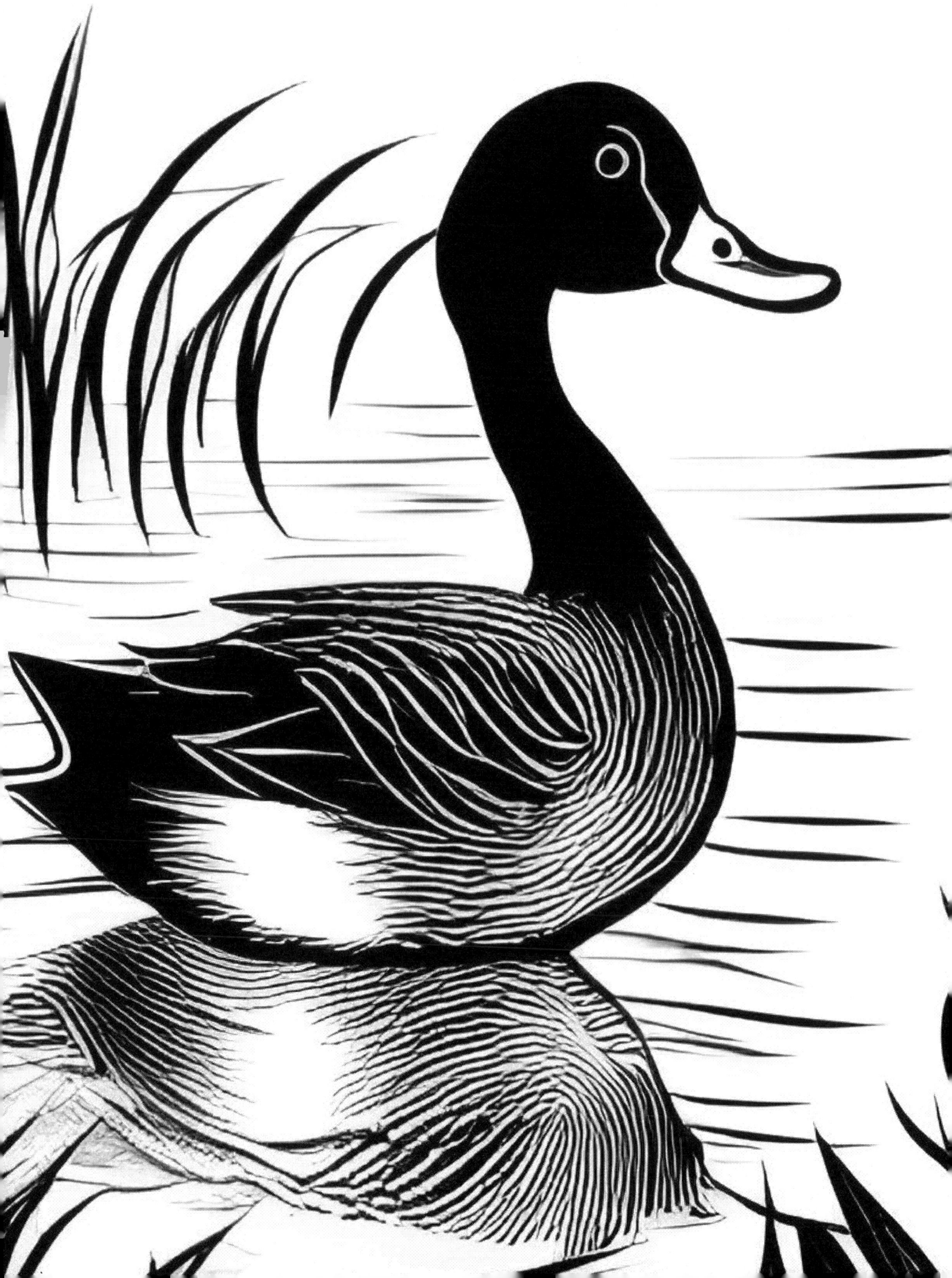

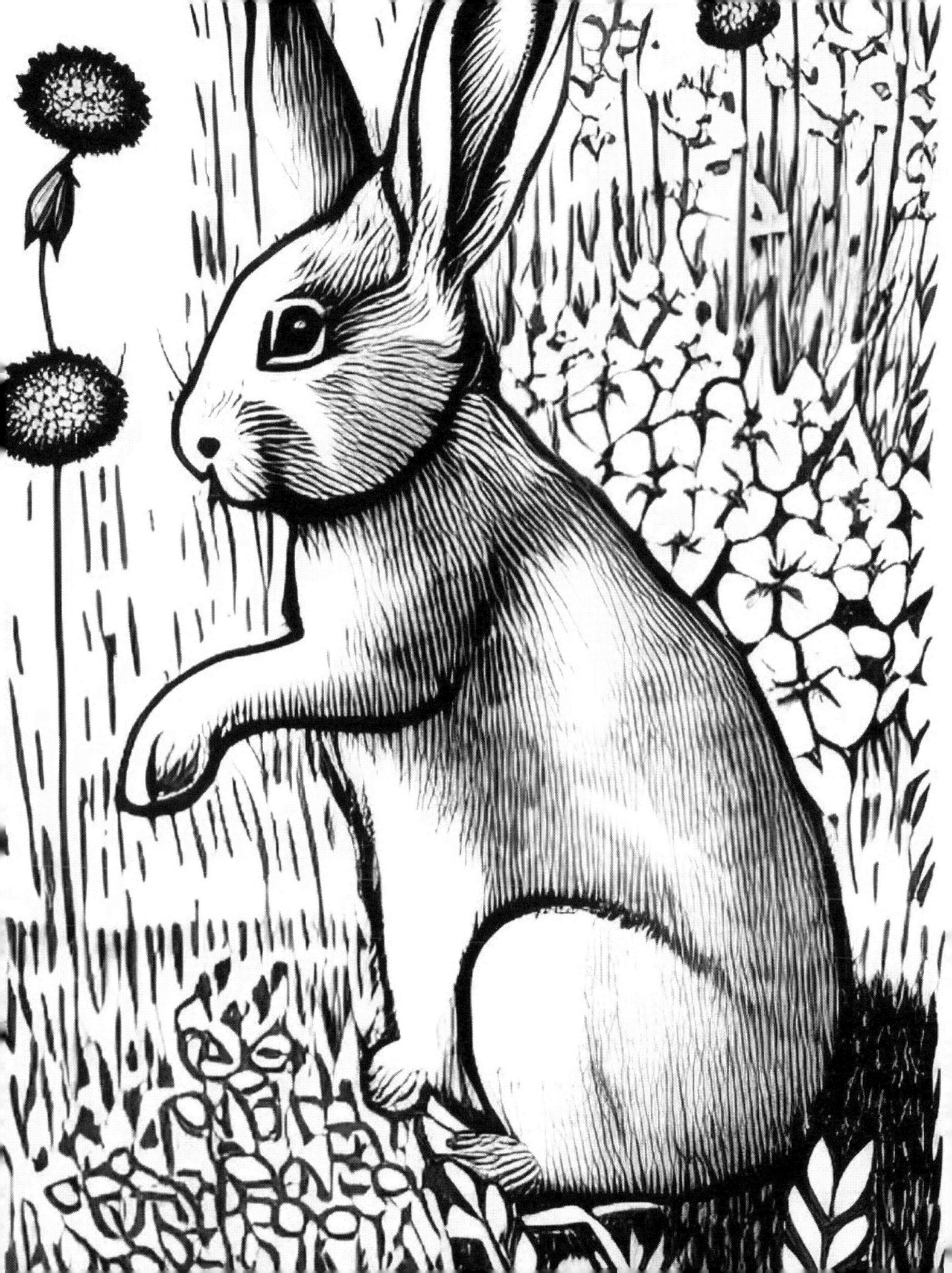

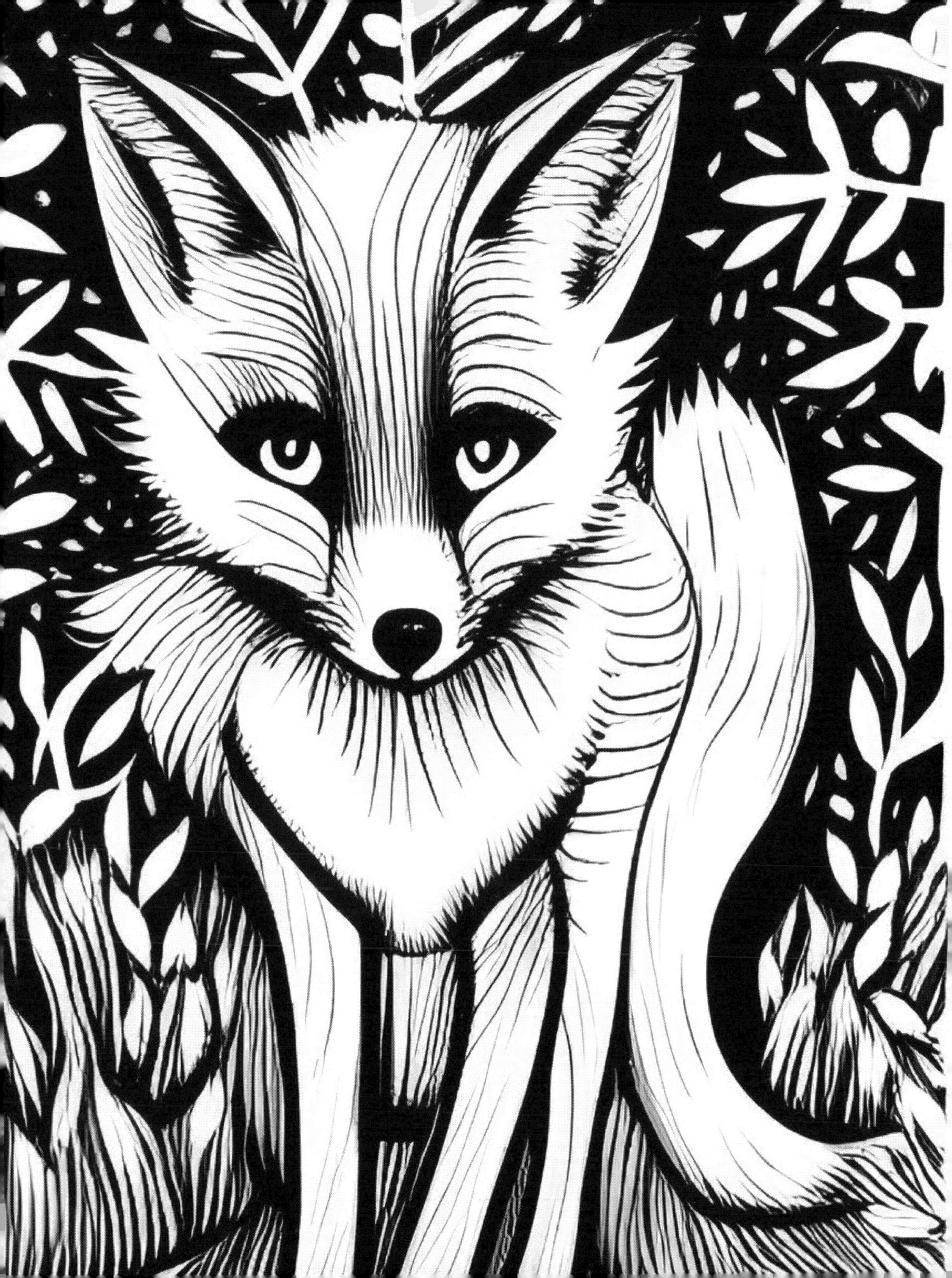

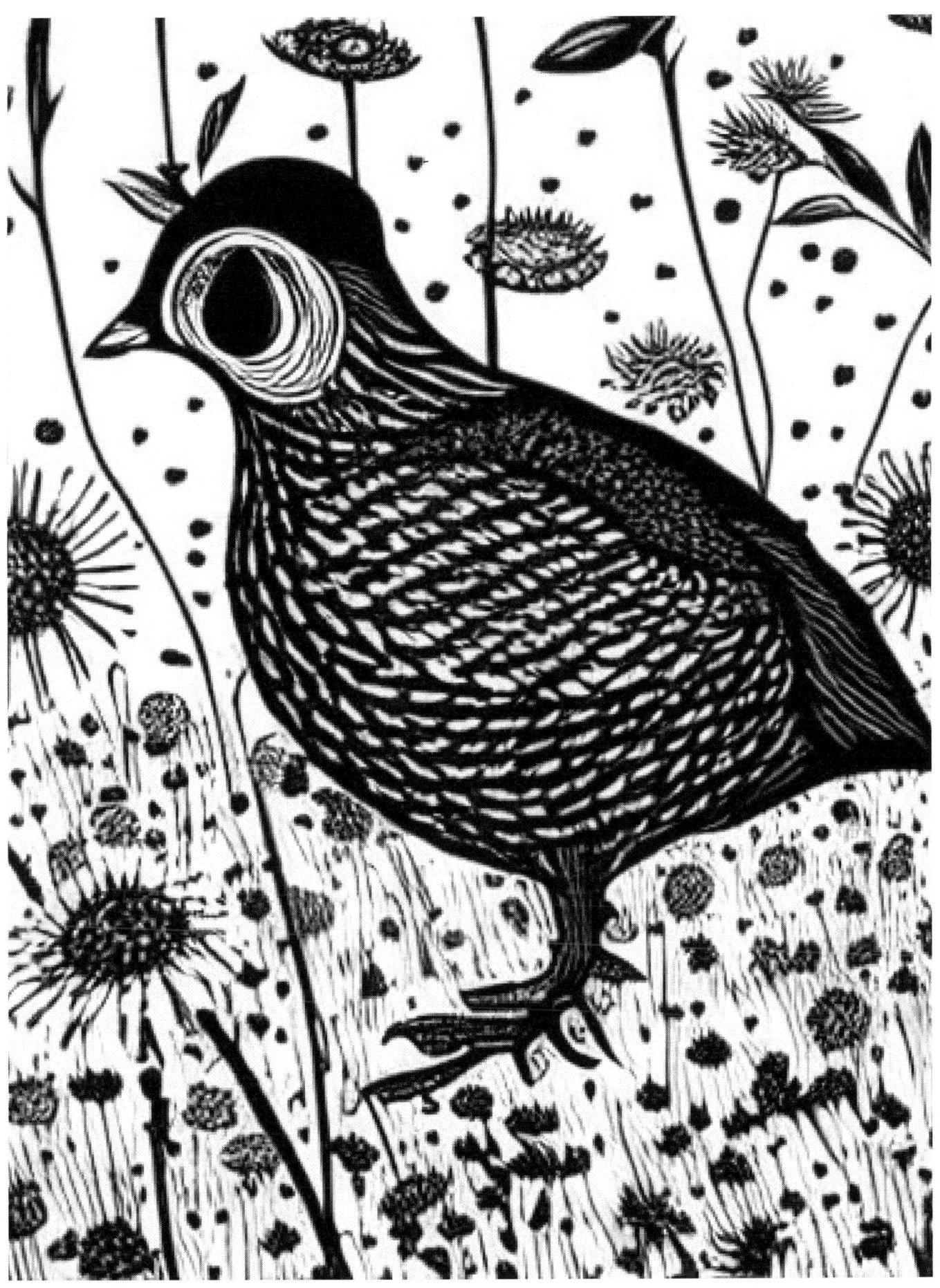

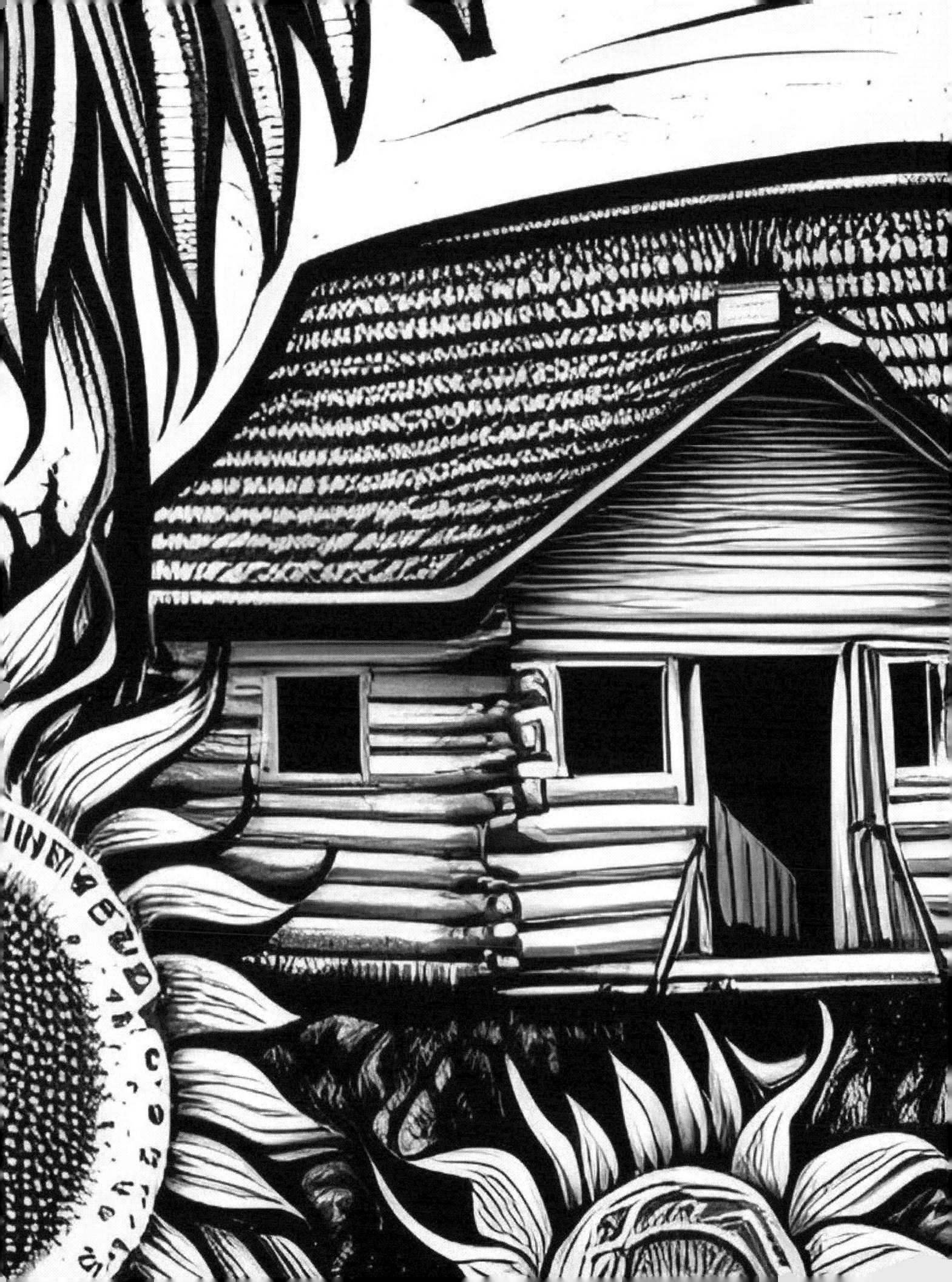

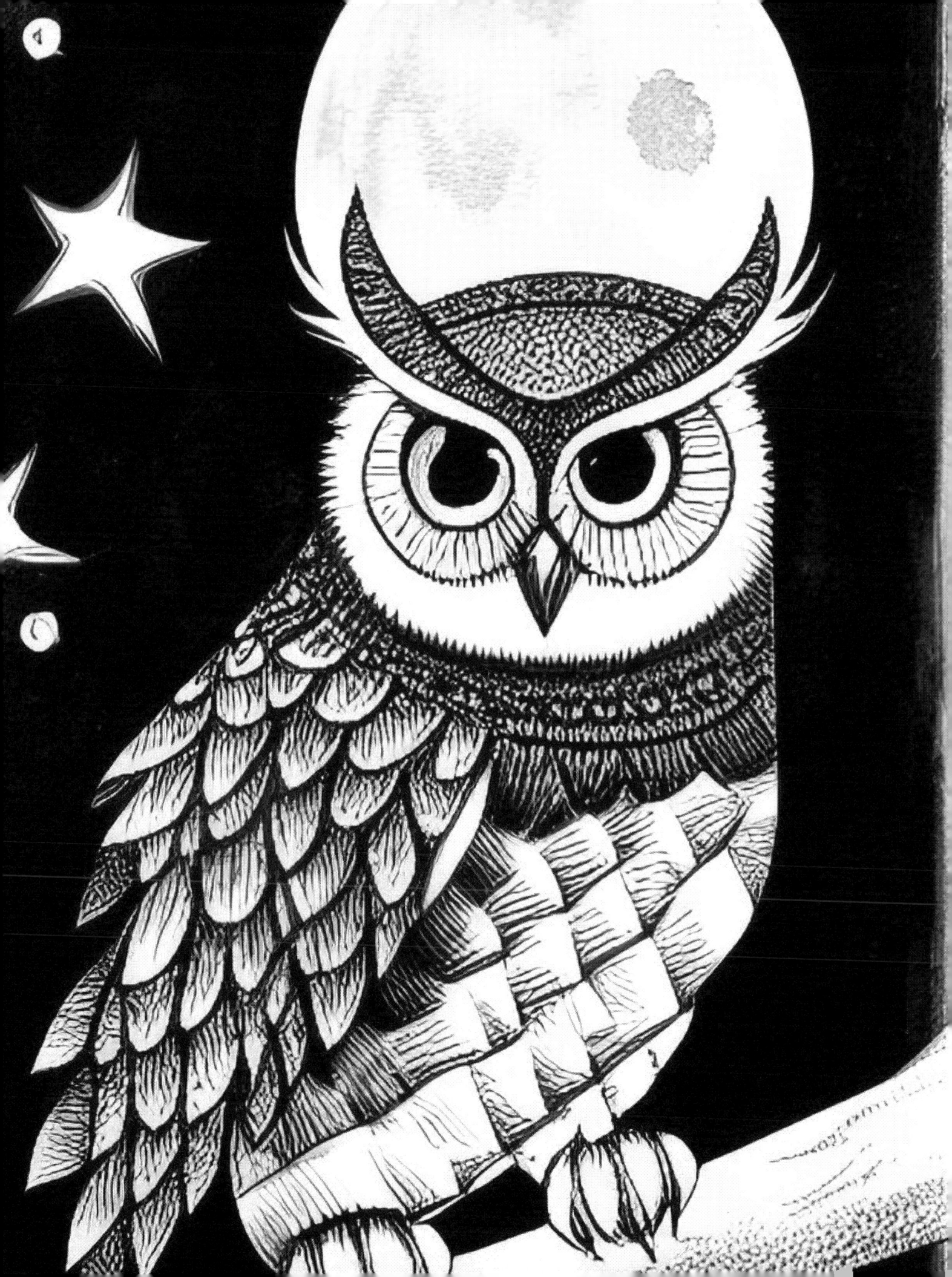

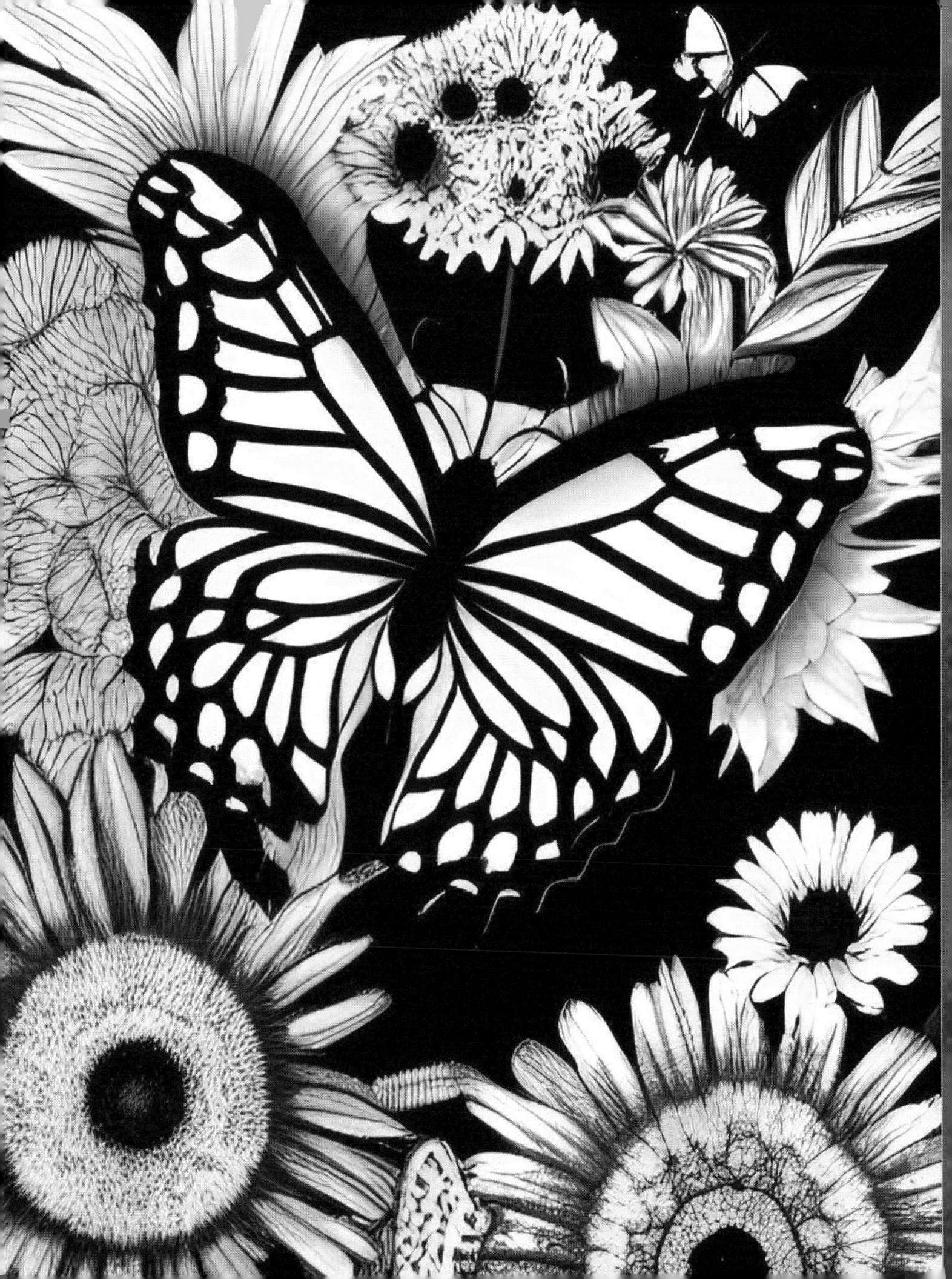

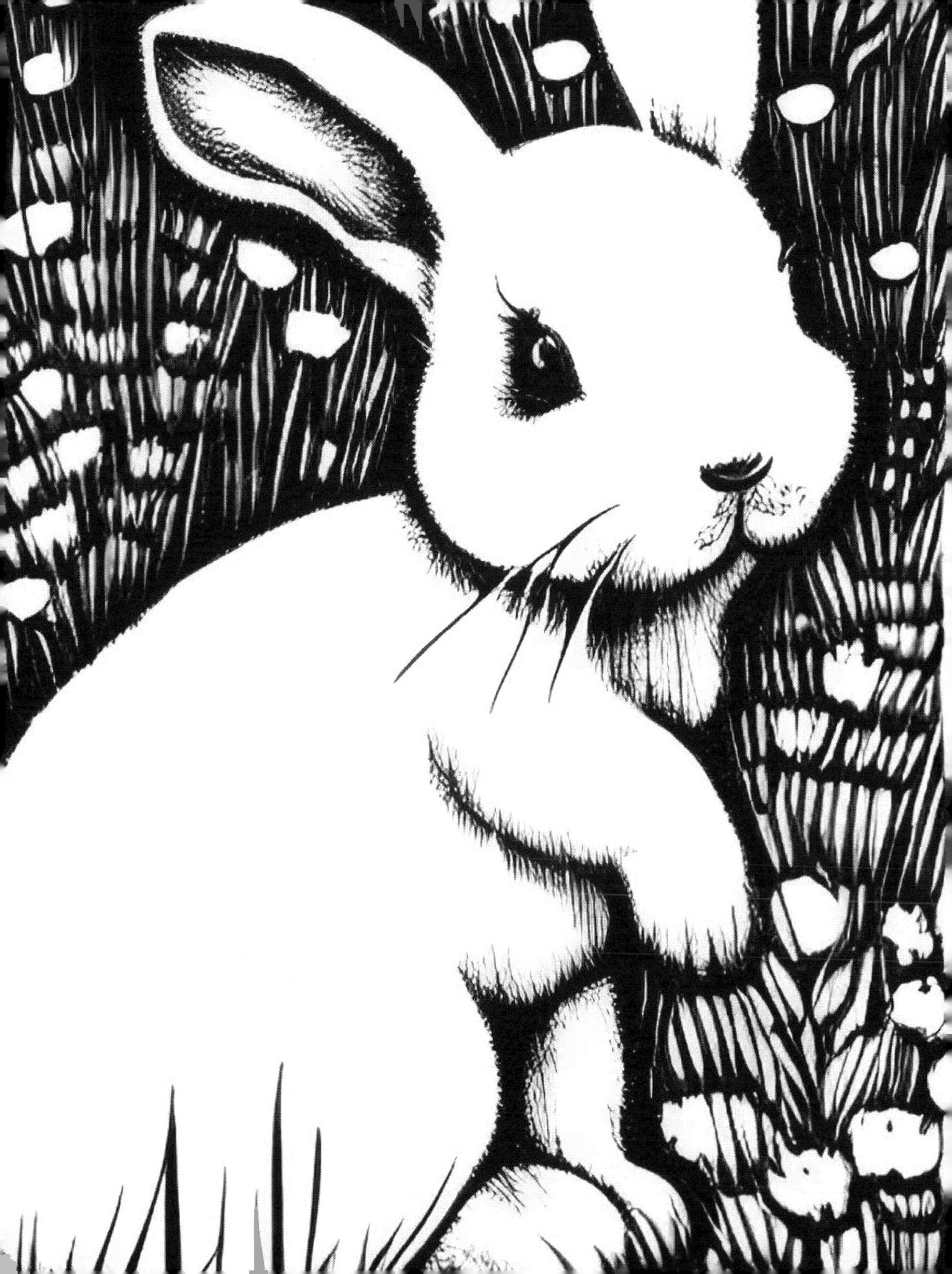

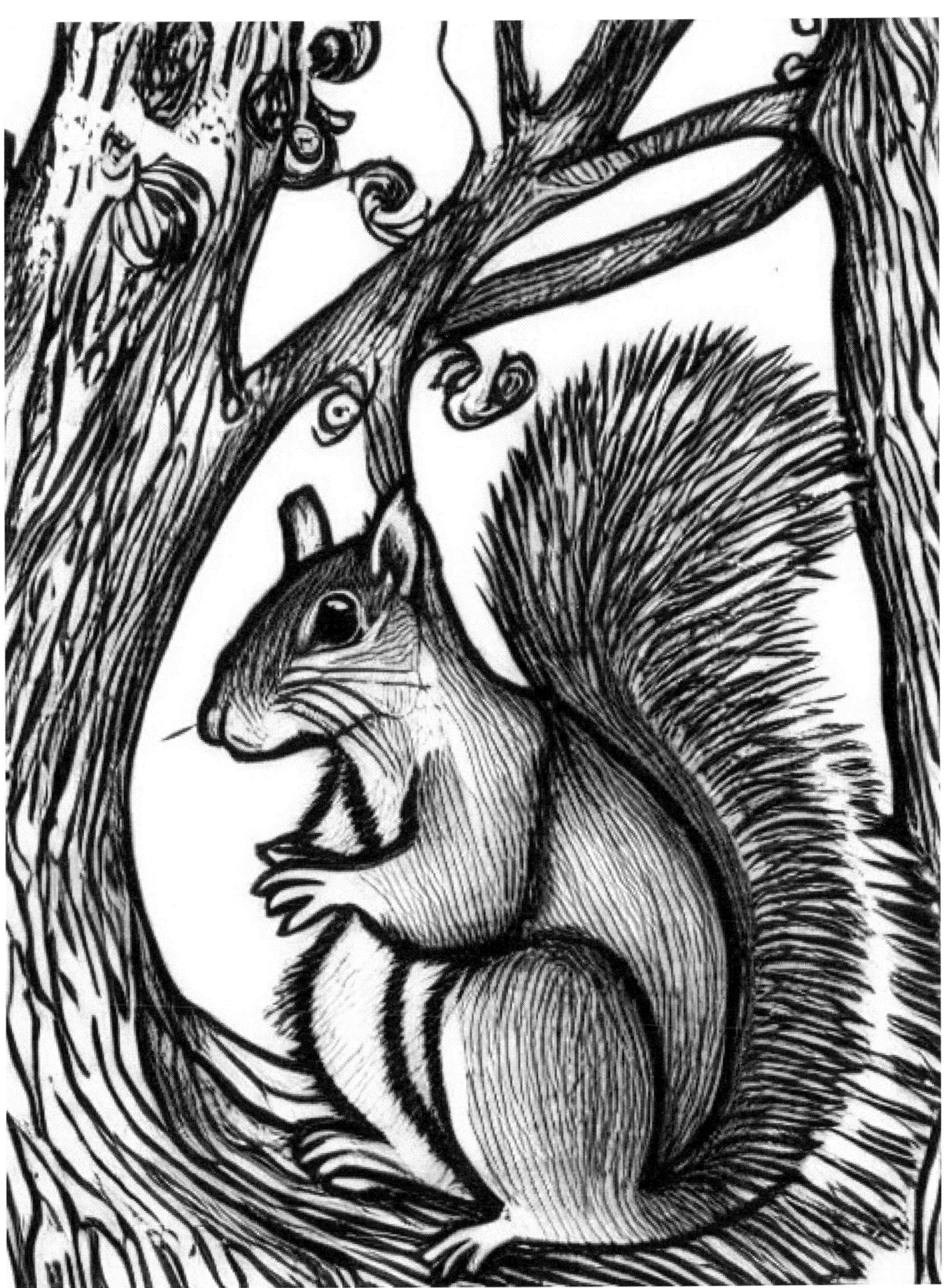

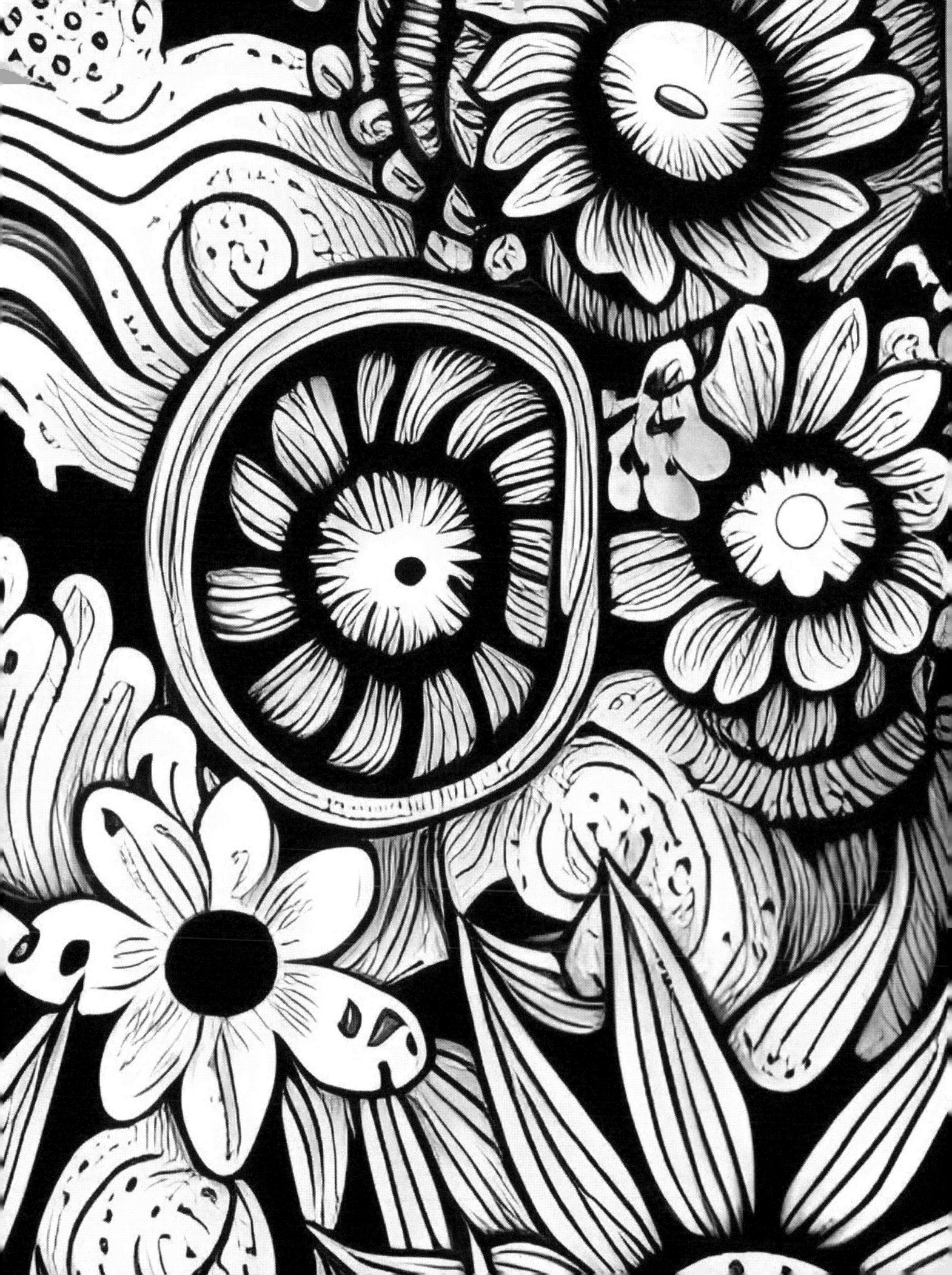

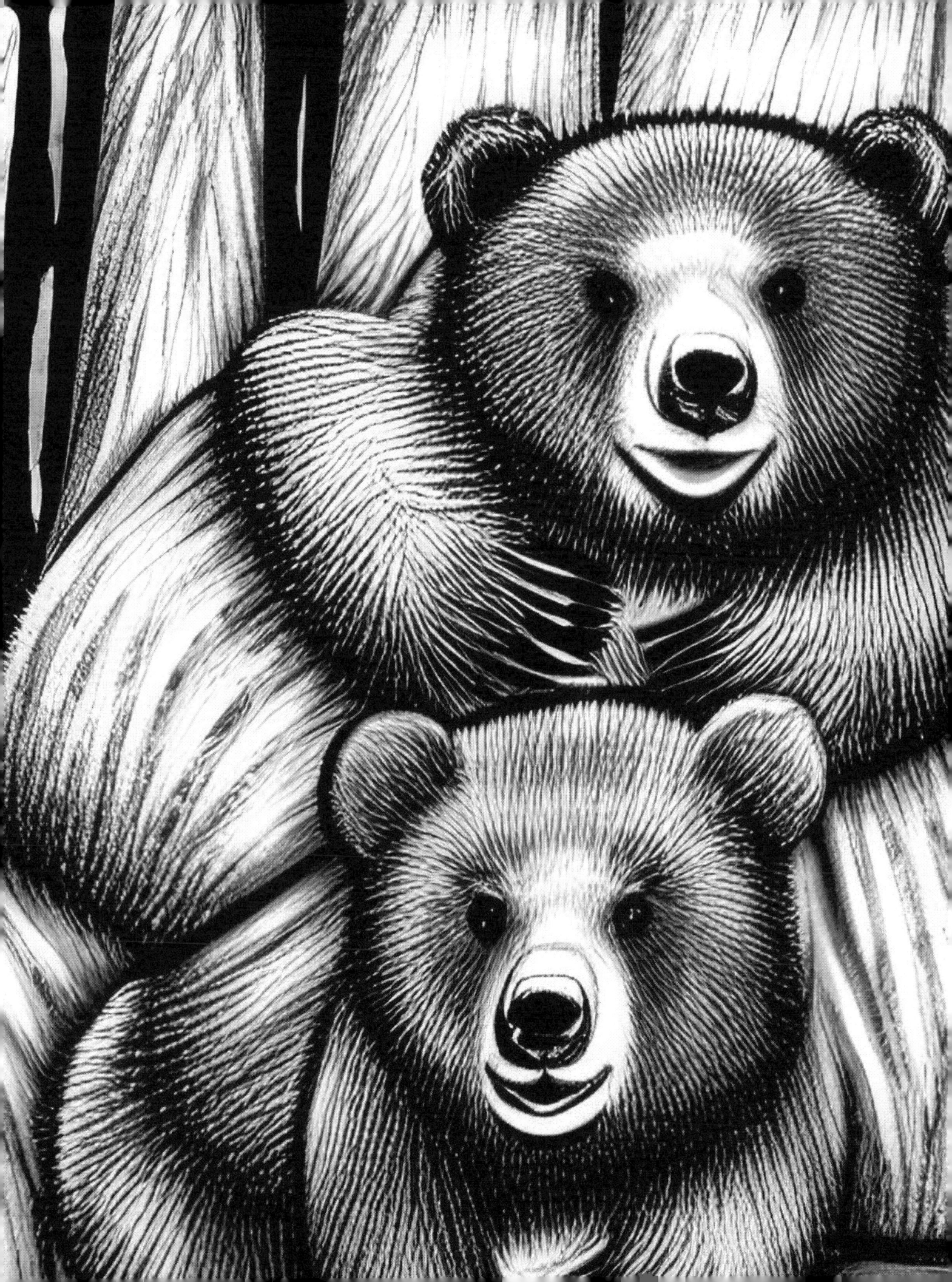

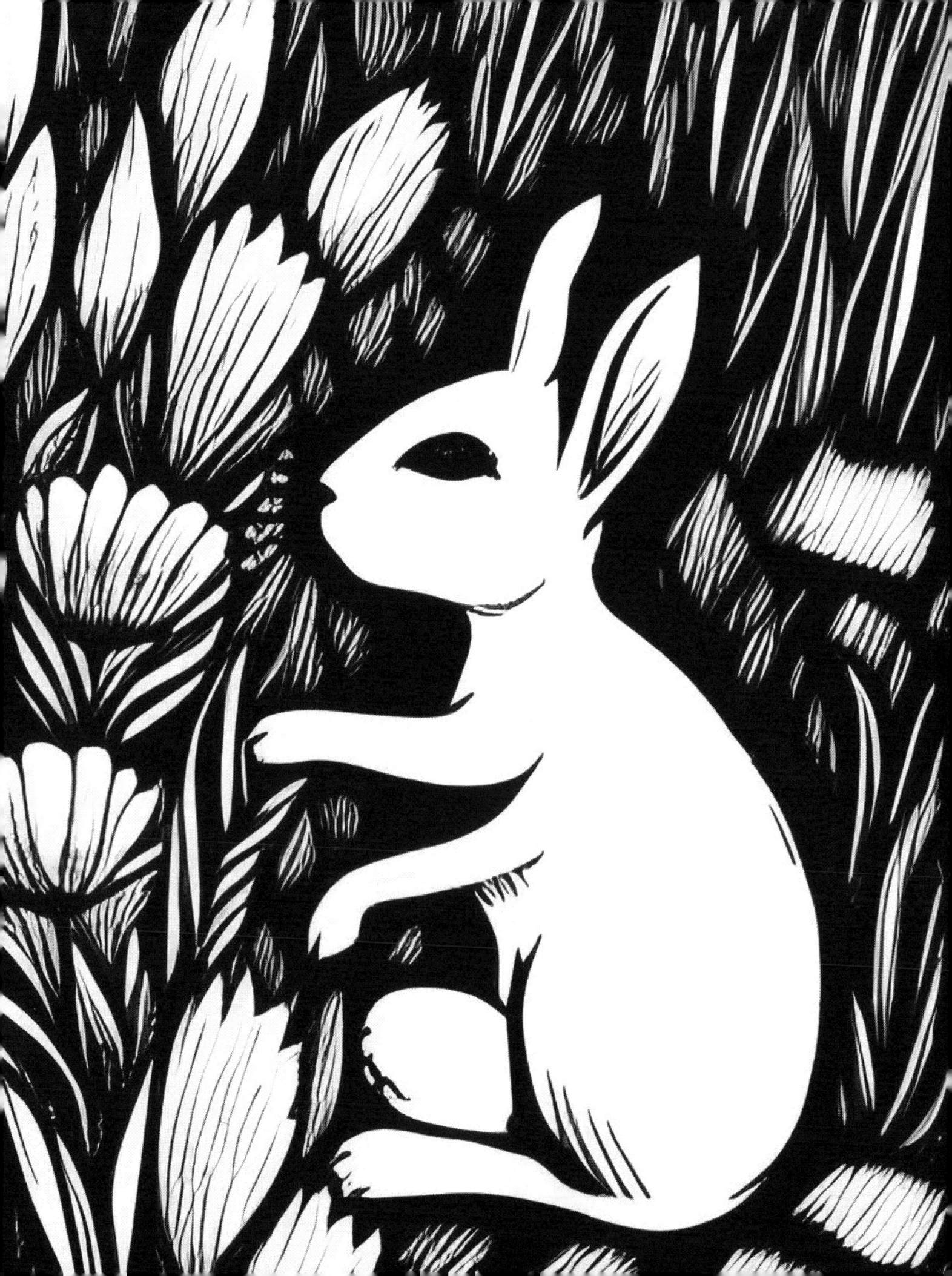

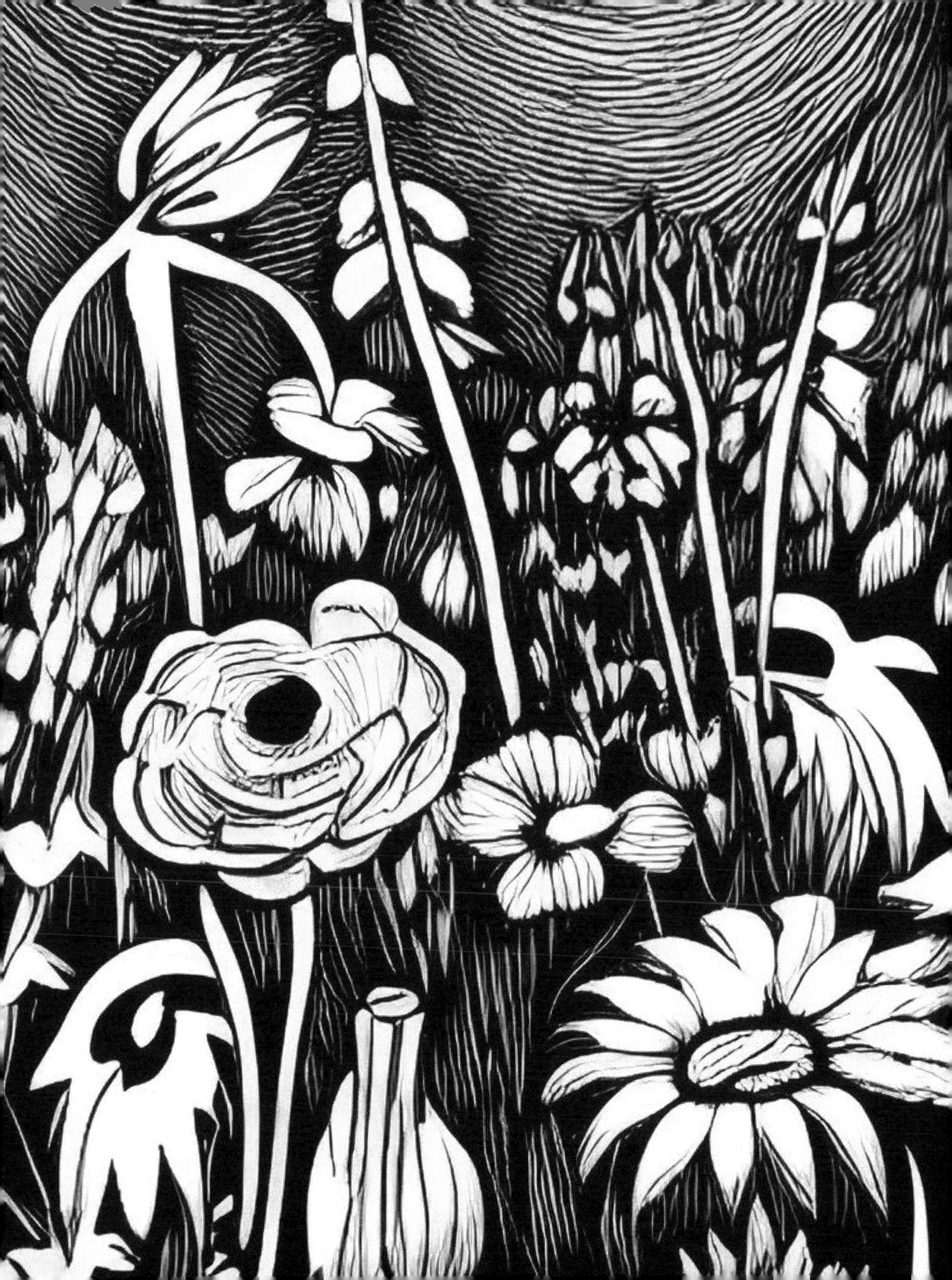

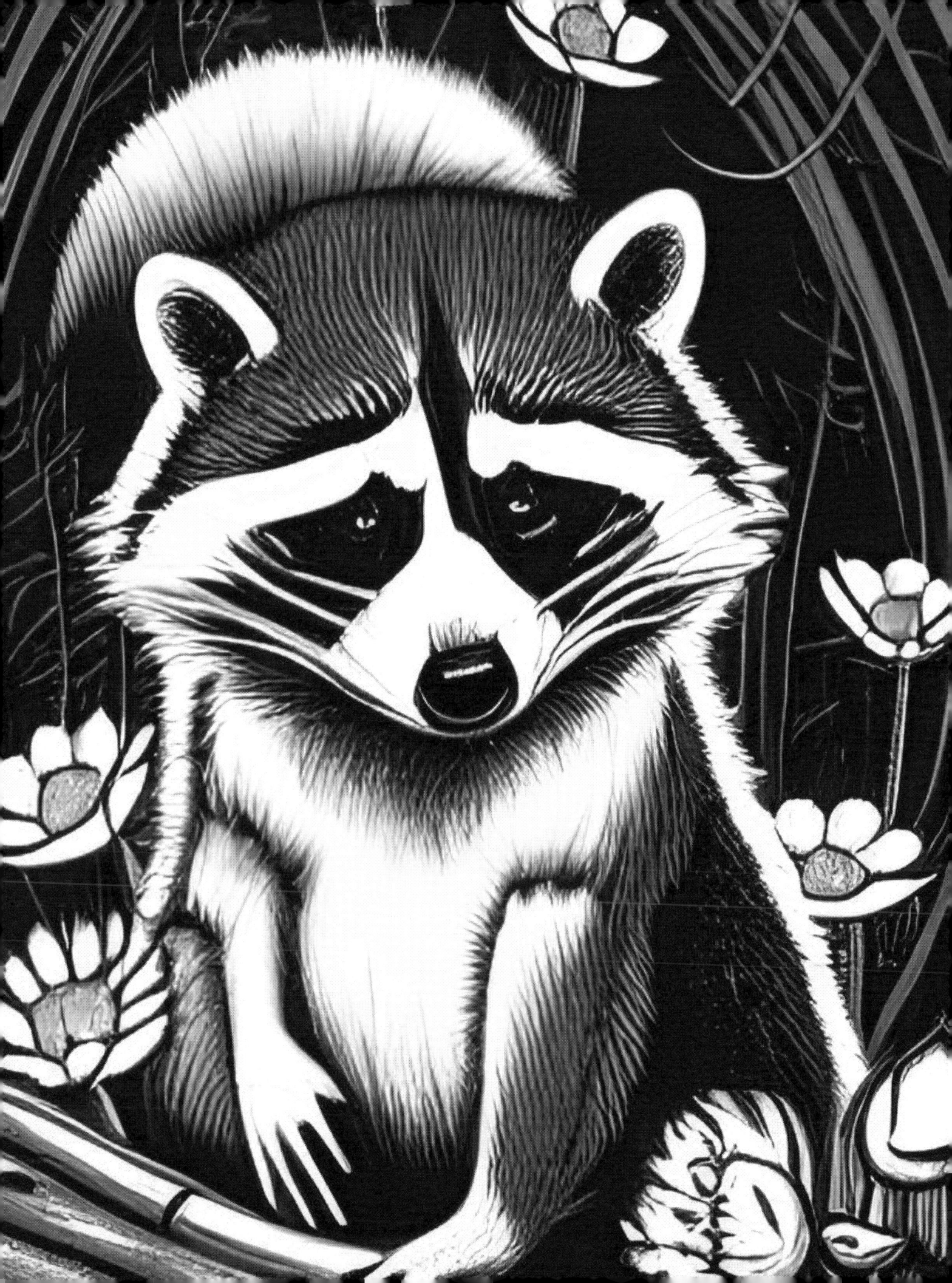

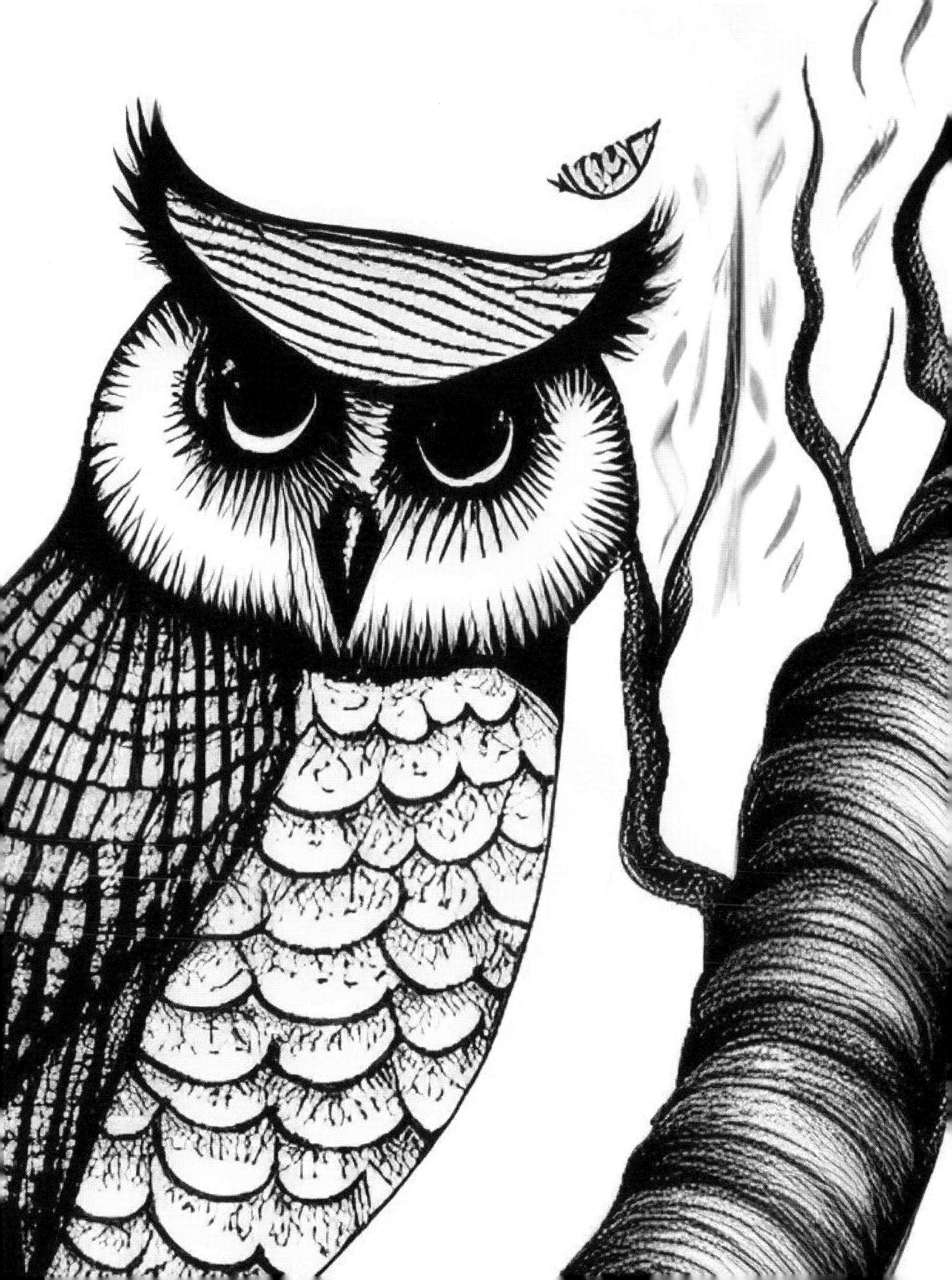

Made in the USA
Columbia, SC
26 April 2023